Machine Vision and
Advanced Image Processing
in Remote Sensing

Springer
Berlin
Heidelberg
New York
Barcelona
Hong Kong
London
Milan
Paris
Singapore
Tokyo

I. Kanellopoulos · G. G. Wilkinson
T. Moons (Eds.)

Machine Vision and Advanced Image Processing in Remote Sensing

Proceedings of Concerted Action MAVIRIC
(Machine Vision in Remotely Sensed Image Comprehension)

With 165 Figures
and 13 Tables

Springer

Ioannis Kanellopoulos
Joint Research Centre, Commission of the European Communities
Space Applications Institute, Environment and Geo-Information Unit
I-21020 Ispra (Varese)
Italy

Prof. Graeme G. Wilkinson
Kingston University
School of Computer Science and Electronic Systems
Penrhyn Road
Kingston Upon Thames, KT1 2EE
United Kingdom

Dr. Theo Moons
Katholieke Universiteit Leuven
Department of Electrotechnical Engineering (ESAT)
Centre for Processing Speech and Images (PSI)
Kardinaal Mercierlaan 94
B-3001 Heverlee
Belgium

ISBN-13: 978-3-642-64260-9   e-ISBN-13: 978-3-642-60105-7
DOI: 10.1007/978-3-642-60105-7

Library of Congress Cataloging-in-Publication Data
Die Deutsche Bibliothek – CIP-Einheitsaufnahme
Machine vision and advanced image processing in remote sensing: proceedings of concerted action MAVIRIC (machine vision in remotely sensed image comprehension); with 13 tables / I. Kanello-poulos ... (ed.). – Berlin; Heidelberg; New York; Barcelona; Hong Kong; London; Milan; Paris; Singapore; Tokyo: Springer, 1999
  ISBN-13: 978-3-642-64260-9

This work is subject to copyright. All rights are reserved, whether the whole or part of the material is concerned, specifically the rights of translation, reprinting, reuse of illustrations, recitation, broadcasting, reproduction on microfilm or in any other way, and storage in data banks. Duplication of this publication or parts thereof is permitted only under the provisions of the German Copyright Law of September 9, 1965, in its current version, and permission for use must always be obtained from Springer-Verlag. Violations are liable for prosecution under the German Copyright Law.

© Springer-Verlag Berlin · Heidelberg 1999
Softcover reprint of the hardcover 1st edition 1999

The use of general descriptive names, registered names, trademarks, etc. in this publication does not imply, even in the absence of a specific statement, that such names are exempt from the relevant protective laws and regulations and therefore free for general use.

Hardcover-Design: Erich Kirchner, Heidelberg

SPIN 10701323   42/2202-5 4 3 2 1 0

Foreword

Since 1994, the European Commission has undertaken various actions to expand the use of Earth observation (EO) from space in the Union and to stimulate value-added services based on the use of Earth observation satellite data. By supporting research and technological development activities in this area, DG XII responded to the need to increase the cost-effectiveness of space-derived environmental information. At the same time, it has contributed to a better exploitation of this unique technology, which is a key source of data for environmental monitoring from local to global scale.

MAVIRIC is part of the investment made in the context of the Environment and Climate Programme (1994-1998) to strengthen applied techniques, based on a better understanding of the link between the remote sensing signal and the underlying bio- geo-physical processes. Translation of this scientific know-how into practical algorithms or methods is a priority in order to convert more quickly, effectively and accurately space signals into geographical information. Now the availability of high spatial resolution satellite data is rapidly evolving and the fusion of data from different sensors including radar sensors is progressing well, the question arises whether existing machine vision approaches could be advantageously used by the remote sensing community. Automatic feature/object extraction from remotely sensed images looks very attractive in terms of processing time, standardisation and implementation of operational processing chains, but it remains highly complex when applied to natural scenes.

The MAVIRIC concerted action provides an international forum gathering numerous acknowledged European experts from both the machine vision and the remote sensing communities. Recent assessment of the state-of-the-art on the application of machine vision or visual processing techniques in remote sensing has been made during a highly successful workshop held in May 1998. The present volume summarises the outcome of this workshop, and draws up research priorities in this area that could lead to new remote sensing applications and market segments. I believe it represents a very significant contribution to this field.

Michel Schouppe
Environment & Climate Programme
DG XII: Science, research and development
European Commission

Table of Contents

Foreword v

Introduction 1

Part I. Image Processing and Computer Vision Methods for Remote Sensing Data

---

Recent Developments in Remote Sensing Technology and the Importance of Computer Vision Analysis Techniques
Graeme G. Wilkinson 5

Posing Structural Matching in Remote Sensing as an Optimisation Problem
Peter Forte and Graeme A. Jones 12

Detail-Preserving Processing of Remote Sensing Images
Silvana Dellepiane 23

Multi-Channel Remote Sensing Data and Orthogonal Transformations for Change Detection
Allan Aasbjerg Nielsen 37

Aspects of Multi-Scale Analysis for Managing Spectral and Temporal Coverages of Space-Borne High-Resolution Images
Bruno Aiazzi, Luciano Alparone, Alessandro Barducci, Stefano Baronti, Roberto Carlà, and Ivan Pippi 49

Structural Inference Using Deformable Models
Jens Michael Carstensen, Rune Fisker, Nette Schultz, and Torsten Dörge 61

Terrain Feature Recognition Through Structural Pattern Recognition, Knowledge-Based Systems, and Geomorphometric Techniques
Dimitris P. Argialas 72

viii   Table of Contents

Part II. High Resolution Data

Environmental Mapping Based on High Resolution Remote Sensing Data
Kolbeinn Arnason and Jon Atli Benediktsson 89

Potential Role of Very High Resolution Optical Satellite Image Pre-Processing for Product Extraction.
P. Boekaerts, V. Christopoulos, A. Munteanu, and J. Cornelis 100

Forestry Applications of High Resolution Imagery
Tuomas Häme, Mikael Holm, Susanna Rautakorpi, and Eija Parmes ... 111

Image Analysis Techniques for Urban Land Use Classification. The Use of Kernel Based Approaches to Process Very High Resolution Satellite Imagery
Charalambos C. Kontoes 121

Part III. Visualisation, 3D and Stereo

Automated Change Detection in Remotely Sensed Imagery
Joseph Mundy and Rupert Curwen 137

A 3-Dimensional Multi-View Based Strategy for Remotely Sensed Image Interpretation
Theo Moons, David Frère, and Luc Van Gool 148

3D Exploitation of SAR Images
Regine Bolter and Axel Pinz 160

Visualizing Remotely Sensed Depth Maps using Voxels
Nilo Stolte 170

Three Dimensional Surface Registration of Stereo Images and Models from MR Images
Henning Nielsen, Lasse Riis Østergaard, and David Le Gall 181

Exploring Multi-Dimensional Remote Sensing Data with a Virtual Reality System
Walter Di Carlo 189

Part IV. Image Interpretation and Classification

Information Mining in Remote Sensing Image Archives
Mihai Datcu, Klaus Seidel, and Gottfried Schwarz 199

Fusion of Spatial and Temporal Information for Agricultural Land Use Identification – Preliminary Study for the VEGETATION Sensor
Jean-Paul Berroir 213

Rule-based Identification of Revision Objects in Satellite Images
Camilla Mahlander and Dan Rosenholm 219

Land Cover Mapping from Optical Satellite Images Employing Subpixel Segmentation and Radiometric Calibration
Werner Schneider 229

Semi-Automatic Analysis of High-Resolution Satellite Images
Wim Mees and Marc Acheroy 238

Density-Based Unsupervised Classification for Remote Sensing
Cees H. M. van Kemenade, Han La Poutré, and Robert J. Mokken 248

Classification of Compressed Multispectral Data
Frank Tintrup, Cristina Perra, and Gianni Vernazza................. 259

Part V. Segmentation and Feature Extraction

Detection of Urban Features Using Morphological Based Segmentation and Very High Resolution Remotely Sensed Data
Martino Pesaresi and Ioannis Kanellopoulos 271

Non-Linear Line Detection Filters
Isabelle Gracia and Maria Petrou 285

Fuzzy Clustering and Pyramidal Hough Transform for Urban Features Detection in High Resolution SAR Images
Eugenio Costamagna, Paolo Gamba, Giacomo Sacchi, and Pietro Savazzi295

x Table of Contents

Detecting Nets of Linear Structures in Satellite Images
Petia Radeva, Andres Solé, Antonio M. López, and Joan Serrat 304

Satellite Image Segmentation Through Rotational Invariant Feature Eigenvector Projection
Albert Pujol, Andrés Solé, Daniel Ponsa, Javier Varona, and Juan José Villanueva 317

Supervised Segmentation by Region Merging
Ben Gorte 328

Introduction

Satellites have been observing the surface of planet Earth for almost thirty years. During that time, sensor systems have steadily improved and become more sophisticated in a number of respects. One of the most important trends as sensors have moved from one generation to another, has been the gradual improvement in ground resolution of captured imagery. As resolution has improved, more and more detail on the land surface has emerged from "blurred" pixels. Ground resolution directly controls the amount of detail, which can be perceived in images, and the precision of cartographic or statistical information which can be extracted.

Within the last two years, the resolution of routinely and openly available remotely sensed images has passed through the 10m. barrier. Imagery can now be acquired with 5m. panchromatic resolution. Within the next 1–2 years, panchromatic imagery with 1m. ground resolution and multispectral imagery with 4m. ground resolution will be available. With such resolution, the structures of many man-made objects become apparent. There are many applications of Earth Observation which will benefit significantly from such high resolution, especially large scale topographic mapping and applications which require detailed information about human activity. In most cases, these new applications will require objects to be identified on the basis of *structure*, *form*, and *context*, as well as by the rather more traditional approach of spectral characteristics. The analysis of imagery to extract structures requires understanding and comprehension based on models and prior knowledge. Such approaches are related to visual perception. It is no surprise therefore to discover that more and more analysis techniques from the world of *computer vision* are beginning to be adopted in remote sensing.

There is an emerging need to bring the latest research ideas from computer vision into remote sensing and also to expose the computer vision community to the new kinds of imagery which will be provided in the near future from space. With this aim in mind, a Concerted Action was established within the European Commission's *Environment and Climate* research programme to explore the link between computer vision and remote sensing. Entitled "MAVIRIC" (*MAchine Vision In Remotely sensed Image Comprehension*), this concerted action brought together experts from the computer vision and remote sensing communities at a workshop, held at Kingston upon Thames, UK, in May 1998. Although primarily a European event, specially invited researchers from outside the European Union countries also participated. The aim of this workshop was to review the current state of the art in the application of computer vision algorithms in remote sensing and to consider future directions for research.

2   Introduction

This volume contains chapters based on the contributions at the Kingston workshop. By presenting them in this form, it is hoped that the ideas of the assembled experts will reach a wider audience. There could indeed be much to be gained by this as the whole future of remote sensing could depend on the success with which advanced vision algorithms are able to extract information from imagery hitherto thought impossible. The challenge is very great indeed.

The chapters in this book have all been edited to ensure consistency in style and approach as far as possible. The editors are very grateful to the contributors, both for their excellent presentations at the workshop itself, and for the very high standard of written material they have prepared subsequently. It is hoped that this book will be a milestone in the documentation of advances in remote sensing, in the same way that the new very high resolution satellites represent a milestone in the history of remote sensing technology. The chapters in this book demonstrate very clearly that significant approaches are being made in remote sensing applications using advanced pattern recognition approaches especially in the urban context. They also show, however, that the reliable extraction of shapes and forms is still problematical and that the modelling of objects and structures still requires much research. It is to be hoped that this volume will stimulate a wider group of scientists to examine these important problems.

The editors would particularly like to thank Nina O'Shea and Paul Giaccone of the School of Computer Science and Electronic Systems at Kingston University, both for their excellent organization of the scientific workshop and for assistance with the preparation of this volume. We also wish to thank Karen Fullerton, Michalis Petrakos and Martino Pesaresi of the Space Applications Institute, at the Joint Research Centre of the European Commission, for their assistance in editing the chapters in this book. Finally we wish to gratefully acknowledge the support of DGXII of the European Commission for the "MAVIRIC" Concerted Action (contract no. ENV4–CT97–0517).

The Editors
November, 1998.

Part I

Image Processing and Computer Vision
Methods for Remote Sensing Data

Recent Developments in Remote Sensing Technology and the Importance of Computer Vision Analysis Techniques

Graeme G. Wilkinson

School of Computer Science and Electronic Systems, Kingston University, Penrhyn Road, Kingston upon Thames, KT1 2EE, UK.
e-mail: G.Wilkinson@kingston.ac.uk

Summary. Remote sensing imagery with "very high" spatial resolution of 10m or less is becoming routinely available. Such imagery permits large scale topographic mapping and land use monitoring. The extraction of meaningful information can be aided by techniques from computer vision. Much research is needed on this problem. It is essential to constrain applications and analysis techniques to those that are solvable and computable as indicated by past experiences in the computer vision field.

1. Introduction

There have been several significant evolutionary steps in satellite remote sensing technologies during the 1990's. These include the development of radar sensors, imaging spectrometers, very high resolution sensors, and multi-view angle sensors. Some of these different types of sensors have already been flown on Earth observation satellites; others will be launched in the next 2-3 years. Collectively these new technologies will provide a significantly enhanced capability for environmental monitoring besides creating new applications for remote sensing. Whilst the original focus of much remote sensing activity was the natural environment, more recently the monitoring of human activities and human impacts on the landscape have become progressively more important.

2. Topographic Mapping from Remote Sensing

Although satellite remote sensing may be used in limited circumstances for map revision, it has never been possible in the past to use it routinely for large scale topographic mapping, nor for the derivation of accurate information concerning land use in zones of dense human habitation. In 1991 it was stated in a review article that "satellite remote sensing has little place in the UK for new topographic mapping" [9]. In the same article, the author also stated that "5m pixel size and stereo capability would significantly increase interpretative capability...the interpretation of complex features is still a

6   G. Wilkinson

highly skilled manual procedure and is likely to remain so until there are significant developments in artificial intelligence and machine vision" [9].

Indeed, possibly the largest land cover mapping exercise ever undertaken with remote sensing, the European Commission's CORINE Land Cover Project for Europe (carried out during the 1980's and early 1990's) relied almost entirely on a labour intensive human photo-interpretation approach. Although recognised as being the only effective approach at the time, subjectivity and lack of reproducibility for statistical purposes have been well noted, e.g. [24].

However, most of the elements stated as requirements for *automated* large scale topographic mapping from space in 1991 are now effectively in place such as the availability of images with 5m pixels, stereo (and multi-view angle) imaging, and major developments in machine vision. However, whilst such elements may exist, much work is still needed on their integration in order to obtain maximum benefit from recent and forthcoming Earth observation satellites. Furthermore, significant economic decisions, such as the allocation of structural funds within Europe, can require timely and accurate statistical information about land use, especially in urban areas. The European Statistical office (EUROSTAT), for example, has recently supported several experimental projects for urban land use mapping from remote sensing. Such projects are now focusing on the use of very high resolution mapping in order to satisfy the needs both of national and EC agencies. At a more mundane level, there is little doubt that high accuracy land use information derived from space can have significant benefits for industry and commerce. For example, the siting of base station aerials for mobile telecommunications can be improved through the use of up-to-date, highly accurate, land use and topography information. Satellite remote sensing is undoubtedly the most reliable and cost-effective way of producing such information for large areas of territory.

3. Very High Resolution Sensors

Very high spatial resolution imaging usually refers to satellite images gathered with pixel resolution of 10m or better. A number of civilian satellites capable of acquiring such images have been developed during the late 1990's representing a new trend in satellite remote sensing. The origins of this technology lie with earlier military surveillance satellites. Very high resolution images are ideal for large scale mapping (e.g. 1:10,000 scale) and for detection and analysis of man made structures on the landscape. This technology has much promise, although legal and ethical issues concerned with privacy are beginning to arise [20].

Several very high resolution remote sensing satellite systems are in development by commercial organisations because of the economic potential of

Recent Developments in Remote Sensing Technology

Table 3.1. High resolution sensors.

Satellite	Main imaging sensor	Spectral bands of sensor (wavelength)	Pixel ground resolution (m)	Date of launch/ expected launch
IRS-1C	Linear Imaging Self-Scanning Sensor (LISS)	Multispectral: 520-590nm 620-680nm 770-860nm 1550-1750nm Panchromatic: 500-750nm	23.5m " " 70.8m 5m	1995
Quickbird	QuickBird Multispectral (QBM)	Multispectral: 450-520nm 530-590nm 630-690nm 770-900nm	4m " " "	4th quarter 1998
	QuickBird Panchromatic (QBP)	Panchromatic: 450-900 nm	1m	
IKONOS-1	Optical Sensor Assembly	Multispectral: 450-530nm 520-610nm 640-720nm 770-880nm Panchromatic: 450-900nm	4m " " " 1m	3rd quarter 1998

detailed land surface observations. Such systems will not gather data continuously because of the very high data rates that would be produced. Instead most very high resolution systems are designed to be "programmable" so that the sensor views a particular chosen ground location as it passes.

Table 3.1 gives specifications of some of the very high resolution remote sensing satellites that should be providing imagery with better than 10m spatial resolution by the end of 1998 [12].

4. Computer Vision Algorithms in Remote Sensing: Current Situation

In general, analysis techniques from machine vision have not so far made a large impact in remote sensing, although there is evidence of a number of techniques being applied in limited experiments. The demarcation between "vision" techniques as opposed to general "image processing" techniques is, in any case, not clear and some techniques can be regarded in some contexts as "vision-based" and in others as "not vision-based". Image processing can be generally understood to include procedures such as image restoration, enhancement and classification – generally requiring low to medium levels of

8 G. Wilkinson

Table 4.1. Machine vision algorithms used in remote sensing to date.

Machine vision algorithm/approach	Use in remote sensing	Example references
Shape analysis	Linear feature detection (roads, drainage patterns, geological lineaments)	[4, 23], [8, 10]
Segmentation	Parcel extraction to aid and improve classification for resource inventories	[16, 19]
Stereo matching	Derivation of digital terrain models	[17]
Optical flow analysis	Monitoring of dynamic phenomena (especially oceanic or atmospheric)	[2, 21]
Data fusion	Integration of images from multiple satellites (esp. optical and radar) to improve landscape classification	[18, 25]
Contextual scene analysis	Integration of ancillary or contextual information for scene classification and mapping	[11]

contextual knowledge. Vision techniques involve scene "understanding" [17] or comprehension, which usually requires models [3, 1], and analysis of objects, structures, or regions – often using artificial intelligence approaches such as expert or knowledge-based systems [6, 15, 22]. It is also possible to regard dynamic scene analysis and fusion of multiple sensor scenes as additional "vision" approaches because of their prevalence in robotic vision applications. Table 4.1 lists some typical vision algorithms or approaches used in remote sensing to date.

In addition to the machine vision techniques listed in table 4.1, some techniques from coherent imaging have begun to be used in remote sensing on account of the availability of coherent synthetic aperture radar imagery. Of particular interest is the technique of interferometry which has been used to detect changes in surface topography (e.g. following earthquakes [14]).

Image understanding and vision techniques have also been applied in the context of the analysis of digital map data [13] and models have also been developed to incorporate image understanding in geographical information systems (GIS) [5]. Such activities extend the application of vision techniques from remote sensing into the closely-related GIS area.

5. Computer Vision Algorithms: Requirements for Success

The advent of very high resolution imagery such as indicated in table 3.1, will undoubtedly lead to the development of a new range of applications for remote sensing particularly in the context of topographic mapping and monitoring of human activity. Spatial resolutions in the region of 1–5m will permit the extraction of very detailed information about man-made objects and infrastructure. However, the automatic interpretation of such scenes will require the development of complex scene and object models following some of the approaches already adopted in robotic and medical imaging. Such approaches have already found some application in the analysis of aerial survey imagery [7]. It will be particularly important, in view of the complexity of the problem arising from the variety of terrestrial landscapes, to constrain visual processing algorithms to realistic and solvable problems. Arguably this has been one of the key elements which has determined whether machine vision applications have succeeded or failed. Indeed the key criteria contributing to successful machine vision applications in the past have generally been the following:

(i) Narrowly constrained application
(ii) Consistent imaging conditions
(iii) Limited number of objects to be detected
(iv) Objects do not vary significantly
(v) Objects are relatively easily modelled

In order to satisfy criteria (i) – (v), it will be important to focus on defining applications of very high resolution remote sensing which are "narrow" and "solvable". It will also be important to concentrate on the generalisation ability of algorithms and models of scenes and objects.

Fortunately, in remote sensing, criterion (ii) is relatively easily met, subject to the imagery being acquired in cloud free conditions. Criterion (iii) is less easily met, as the range of observable features on the Earth's surface is enormous. However, some features can be initially selected by their spectral characteristics (e.g. of concrete) before more sophisticated algorithms to classify on the basis of shape or context are applied.

The actual detection of objects which may be required in topographic mapping or land use analysis will depend very heavily on the variations which might be expected in the appearances of such objects in different contexts (i.e. criterion (iv)). Moreover, there is a close inter-relationship between the extent of variability of such objects and the effectiveness of models built to describe them (criterion (v)). Overall, very high resolution remote sensing will only solve large scale mapping problems if the objects to be detected and the models used to describe them match very precisely with a low level of ambiguity.

Another key issue in automated large scale mapping is likely to be the *computability* of vision algorithms. Techniques which attempt to match highly complex object models to complex multispectral scenes are unlikely to succeed in an operational context because of computation time requirements. Satellite imagery can contain many millions of pixels, which may need to be inspected as spatial groups. The detection and classification of structural objects may involve the analysis of combinatorially explosive combinations of pixel groupings, which may carry significant computational overhead. For this reason, much research will be needed on software architectures to subdivide structural classification problems into separate computable stages. It is likely that object-oriented methods will be highly appropriate in this context. At present such research is only in its infancy in remote sensing.

6. Conclusions

The successful exploitation of the new generation of very high resolution imagery from space will require considerable development of new algorithms based on approaches in machine vision. The construction of useful automatic extraction techniques must follow some of the guiding principles established in the machine vision context. This will include restricting problem definitions and defining target output products to be achievable and computable within the bounds of current structural object modelling techniques.

References

1. D. P. Argialas and C. A. Harlow, "Computational image interpretation models: an overview and a perspective", *Photogrammetric Engineering and Remote Sensing*, vol. 56, no. 6, pp. 871–886, 1990.
2. J.-P. Berroir, S. Bouzidi, I. I. Herlin, and I. Cohen, "Vortex segmentation on satellite oceanographic images", in *Proceedings of the International Society of Optical Engineering*, vol. SPIE 2315, pp. 635–645, 1994.
3. T. O. Binford, "Survey of model-based image understanding systems", *International Journal of Robotics Research*, vol. 1, no. 1, pp. 18–62, 1982.
4. J. Van Cleynenbreugel, F. Fierens, P. Suetens, and A. Oosterlinck, "Delineating road structures on satellite imagery by a GIS-guided technique", *Photogrammetric Engineering and Remote Sensing*, vol. 56, no. 6, pp. 893–898, 1990.
5. M. Gahegan and J. Flack, "A model to support the integration of image understanding techniques within a GIS", *Photogrammetric Engineering and Remote Sensing*, vol. 62, no. 5, pp. 483–490, 1996.
6. D. G. Goodenough, M. Goldberg, G. Plunkett, and J. Zelek, "An expert system for remote sensing", *IEEE Transactions on Geoscience and Remote Sensing*, vol. 25, no. 3, pp. 349–359, 1987.
7. A. Gruen, E. P. Baltsavias, and O. Henricsson (*eds.*), *Automatic extraction of man-made objects from aerial and space images (II)*. Basel: Birkhauser, 1997.
8. F. C. Hadipriono, J. G. Lyon, W. H. Thomas Li, and D. Argialas, "The development of a knowledge-based expert system for analysis of drainage patterns", *Photogrammetric Engineering and Remote Sensing*, vol. 56, no. 6, pp. 905–909, 1990.
9. W. S. Hartley, "Topographic mapping and satellite remote sensing: is there an economic link?", *International Journal of Remote Sensing*, vol. 12, no. 9, pp. 1799–1810, 1991.

10. A. Karnieli, A. Meisels, L. Fisher, and Y. Arkin, "Automatic extraction and evaluation of geological linear features from digital remote sensing data using a hough transform", *Photogrammetric Engineering and Remote Sensing*, vol. 62, no. 5, pp. 525–531, 1996.

11. C. C. Kontoes and D. Rokos, "The integration of spatial context information in an experimental knowledge-based system and the supervised relaxation algorithm – two successful approaches to improving SPOT-XS classification", *International Journal of Remote Sensing*, vol. 17, no. 16, pp. 3093–3106, 1996.

12. H. J. Kramer, *Observation of the Earth and Its Environment*. Berlin: Springer-Verlag, 3rd ed., 1996.

13. L.-H. Lee and T.-T. Su, "Vision-based image processing of digitized cadastral maps", *Photogrammetric Engineering and Remote Sensing*, vol. 62, pp. 533–538, 1996.

14. D. Massonet and F. Adragna, "A full-scale validation of radar interferometry with ERS-1: the landers earthquake", *Earth Observation Quarterly*, vol. 41, pp. 1–5, 1993.

15. D. M. McKeown, "The role of artificial intelligence in the integration of remotely sensed data with geographic information systems", *IEEE Transactions on Geoscience and Remote Sensing*, vol. 25, no. 3, pp. 330–348, 1987.

16. J. Le Moigne and J. C. Tilton, "Refining image segmentation by integration of edge and region data", *IEEE Transactions on Geoscience and Remote Sensing*, vol. 33, no. 3, pp. 605–615, 1995.

17. J.-P. A. L. Muller, "Key issues in image understanding in remote sensing", in *Philosophical Transactions of the Royal Society of London*, vol. 324 of *A*, pp. 381–395, 1998.

18. C. Pohl and J. L. Van Genderen, "Multisensor image fusion in remote sensing: concepts, methods and applications", *International Journal of Remote Sensing*, vol. 19, no. 5, pp. 823–854, 1998.

19. S. Ryherd and C. Woodcock, "Combining spectral and texture data in the segmentation of remotely sensed images", *Photogrammetric Engineering and Remote Sensing*, vol. 62, no. 2, pp. 181–194, 1996.

20. E. T. Slonecker, D. M. Shaw, and T. M. Lillesand, "Emerging legal and ethical issues in advanced remote sensing technology", *Photogrammetric Engineering and Remote Sensing*, vol. 64, no. 6, pp. 589–595, 1998.

21. Y. Sun, "Automatic ice motion analysis from ERS-1 SAR images using the optical flow method", *International Journal of Remote Sensing*, vol. 17, no. 11, pp. 2059–2087, 1996.

22. A. Tailor, A. Cross, D. C. Hogg, and D. C. Mason, "Knowledge-based interpretation of remotely sensed images", *Image and Vision Computing*, vol. 4, no. 2, pp. 67–83, 1986.

23. F. Wang and R. Newkirk, "A knowledge-based system for highway network extraction", *IEEE Transactions on Geoscience and Remote Sensing*, vol. 26, no. 5, pp. 525–531, 1988.

24. G. G. Wilkinson, S. Folving, K. Fullerton, and J. Mégier, "A study on the automatic revision of the European community's CORINE land cover database using satellite data", *International Archives of Photogrammetry and Remote Sensing*, vol. XXIX, pp. 543–548, 1992.

25. G. G. Wilkinson, S. Folving, I. Kanellopoulos, N. McCormick, K. Fullerton, and J. Mégier, "Forest mapping from multi-source satellite data using neural network classifiers – an experiment in Portugal", *Remote Sensing Reviews*, vol. 12, pp. 83–106, 1995.

Posing Structural Matching in Remote Sensing as an Optimisation Problem

Peter Forte and Graeme A. Jones

School of Computer Science & Electronic Systems, Kingston University,
Penrhyn Road, Kingston upon Thames, Surrey KT1 2EE, UK.
e-mail: {P.Forte,G.Jones}@kingston.ac.uk

Summary. The use of computer vision methodologies in remote sensing is likely to rise significantly in the near future with the launch of imaging sensors with very high spatial resolution: offering panchromatic resolutions of between 3 and 1 metres. This will begin to make it possible to use more complex shape models and context information in the analysis of man-made structures. This is likely to revolutionise topographic mapping, cartographic registration, and urban zone monitoring in particular. Extracting extended image features and their relationships from images will enable the application of structural matching techniques to three types of remote sensing registration problems: *image-to-image* registration, *image-to-symbol* cartographic registration, and *multi-modal* registration e.g. matching active SAR data to passively acquired chromatic data. In the rest of this paper we shall review the range of structural matching methods reported in the computer vision literature.

1. Introduction

Satellite-borne imaging sensors now operate in different parts of the electromagnetic spectrum ranging from optical, infra-red and microwave wavelengths. The majority of optical and infra-red sensors are passive collectors of reflected electromagnetic radiation, while in the microwave region, active synthetic aperture radar (SAR) systems transmit polarised radar pulses and measure the backscatter signal. The pixel resolutions of these systems currently range from several meters to several kilometres at the Earth's surface. The spatial resolution of images currently on offer is not sufficient to enable the deployment of the rich vein of machine vision techniques developed in the computer vision community. features. This situation is changing however with imminent increases in resolution as a new generation of satellites become available. Earth observation is currently in an expansive phase with new satellites planned for launch over the next 10 years as a result both of new government programmes such as NASA's Mission to Planet Earth and other commercial programmes. The use of computer vision methodologies in remote sensing is likely to rise significantly in the near future with the launch of imaging sensors with very high spatial resolution: offering panchromatic resolutions of between 3 and 1 metres. This will begin to make it possible to use more complex shape models and context information in the analysis of man-made structures. This is likely to revolutionise topographic mapping, cartographic registration, and urban zone monitoring in particular.

Extracting extended image features and their relationships from images will enable the application of structural matching techniques to three types of remote sensing registration problems: *image-to-image* registration, *image-to-symbol* cartographic registration [3], and *multi-modal* registration [14] e.g. matching active SAR data to passively acquired chromatic data. Although in the rest of this paper, we shall be largely assuming an *image-to-image* problem, the discussion is also applicable to both *image-to-symbol* and *multi-modal* problems. By structural matching we mean the mapping of a model graph structure (nodes for image features and arcs for inter-feature relationships) constructed from one image to a target graph structure constructed in the second.

2. Matching Feature Attributes

The computer vision literature reports systems which attempt to establish stereo correspondence using a large variety of image features such as *edges, points of interest, line segments, contours, regions* and even colour. Low level features such as edges exhibit two main disadvantages as matching tokens. First, the hundreds of edge points extracted demand very large computational resources which precludes the use of exhaustive search strategies. The second problem (which compounds the first) is the paucity of token attribute information which results in very large number of ambiguous candidate matches all of which must be explored. The most common higher level matching tokens employed in recent years have been *line segments*, which are both less numerous than edge points and ideal for constructing relatively rich attribute information such as *length, width, adjacent greylevels* and *contrast* [7, 13]. Other more complex types of token have also been reported: *contours* [9, 15] and *regions* [1, 16].

To continue the introduction to the matching process and sources of constraints, some terms which will be used throughout this section of the report need to be defined. Let λ refer to a feature extracted from the first image and Λ to the set of all features extracted where $\lambda \in \Lambda$. Let ω represent a feature from the second image, and Ω be the set of all features from the second image such that $\omega \in \Omega$. The position of the token λ in the image is given by the vector $\mathbf{u}_\lambda$ *i.e.* the position of line feature may be described by its centroid. The attributes of the token λ are stored in an attribute vector $\mathbf{a}_\lambda = (a_1, a_2, \ldots, a_K)^T$ where K is the number of attribute values. In the same manner, $\mathbf{a}_\omega$ contains a similarly ordered set of attributes for the feature ω). Each feature λ (or ω) has a set of neighbouring features $\mathcal{N}_\lambda$ (or $\mathcal{N}_\omega$) - see figure 2.1(b). Such a local neighbourhood is usually constructed from pairwise *connectedness* or *adjacency* relationships between features.

A match γ is constructed from two features one from each image $\gamma = \{\lambda, \omega\}$. Let $p(\gamma)$ be the probability of the match γ. The matching process may now be defined as a procedure for computing the set of match probabilities

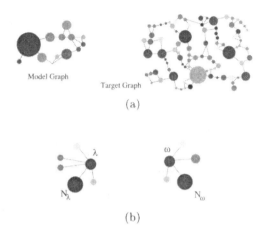

Fig. 2.1. (a) Graph matching (b) Feature neighbourhoods.

and selecting the set of matches whose threshold is greater than some selection threshold τ. The initial probability of each match may be generated from the joint *probability density function* of feature attributes as described below. This *initial candidate match generation* procedure exploits one important source of match constraints - *unary* match constraint. Sources of constraints fall broadly into two categories: *unary* constraints such as attribute similarity which contain information specific to each match. *Binary* constraint measures the *compatibility* of pairs of matches. The set of candidate matches between two images is $\mathcal{M}$ whose worst case size is equal to the outer product of the sets Λ and Ω which for even relatively complex tokens can be large and computationally demanding set to process. In general set of the matches may be pruned by invoking the unary constraint. The second type of constraint is often used to provide match support in the inevitable situations where there are conflicting match interpretations.

2.1 Unary Match Constraint

Match support or inhibition may be derived by comparing each pair of attribute vectors based on the assumption that the viewpoints are similar (or that the relative scaling is known). For each feature's attribute vector, an associated attribute covariance matrix can be recovered. Given two matched feature $\{\lambda, \omega\}$ and their attribute vector $\mathbf{a}_\lambda$ and $\mathbf{a}_\omega$, a dissimilarity measure may be computed from the normalised distance between the attribute vector *i.e.* the *Mahalanobis* distance

$$S(\gamma) = (\mathbf{a}_\lambda - \mathbf{a}_\omega)^T [C_\lambda + C_\omega]^{-1} (\mathbf{a}_\lambda - \mathbf{a}_\omega) \qquad (2.1)$$

where C_λ and C_ω are the covariance matrices of feature λ and ω respectively. Such an approach provides an excellent mechanism for combining and appropriately weighting the contributions of each component of a feature's attributes on the solution.

While the Mahalanobis distance provides a useful measure of the dissimilarity of two features. It is more convenient from a matching (and later optimisation) point of view to convert the Mahalanobis distance into a match probability value. This may be achieved by considering $S(\gamma)$ as a variable drawn from a Chi-Squared distribution of K degrees of freedom. The probability density function $p(S)$ of such a distribution is given by

$$p(S) = \frac{1}{2^{\frac{K}{2}} \Gamma(\frac{K}{2})} S^{\frac{K}{2}-1} \exp\left(-\frac{S}{2}\right) \tag{2.2}$$

where $\Gamma(x)$ is the Gamma function. Thus the initial match probability $p_0(\gamma)$ of a match γ may be estimated from the attributes via equations 2.1 and 2.2.

2.2 Pruning the Initial Match Set

The outer product of two sets of regions Λ and Ω is potentially a very large set. Computing the match probabilities by attribute comparison alone does not usually result in a satisfactory match set. Among the vary large number of false matches, many accidental feature similarities generate relatively high initial match probabilities while noise does adversely affect the initial probabilities of correct matches. In section 4 attribute and structural information is brought together within an optimisation framework to accurately locate the correct match set. These techniques however work most effectively when the number of false matches is at most one, two or three times larger than the number of correct matches. However without further action, the ratio of false matches to correct matches is approximately $\min(|\Lambda|, |\Omega|) : 1$ where $|\cdot|$ denotes the size of a set. Consequently, the huge initial set of matches must be pruned.

While attribute information alone does not discriminate powerfully enough to generate the accurate initial match probability values $p_0(\gamma)$, it is sufficient to partially order the set of matches. In general false matches have small values of probability while correct matches tend to have larger values. Therefore an initial probability pruning can be introduced to exclude the most unlikely matches. The value of the probability pruning threshold must be estimated empirically although a well chosen set of attributes does reduce the sensitivity of the method to this value. The effect of varying threshold on the selected set of matches is demonstrated in figure 2.2 using two plots: the *percentage of correct matches* selected as a function of threshold, and the *ratio of all to correct matches* selected as a function of the threshold. The sketch illustrates highly typical behaviours achieved for well chosen attributes. The *percentage of correct matches* selected is stable for a large range of threshold values. Only at high thresholds do correct matches start to get pruned out. The *ratio of all to correct matches* drops dramatically for modest threshold

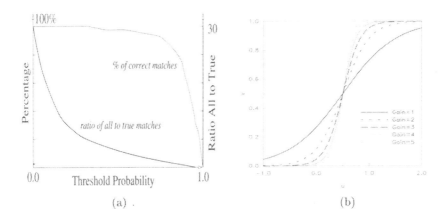

Fig. 2.2. (a) Pruning matches (b) Sigmoidal gain function.

values. Only at high thresholds does the number of both false and correct matches surviving drop dramatically. While pruning still does not prevent the selection of a large number of false matches, the lower proportion should enable the correct matches to be located using more complex optimisation strategies which incorporate structural information.

3. Compatibility between Matches

As previously argued match-specific information is not usually sufficient to allow unambiguous interpretation of the scene. Since the collection of features in both images share similar structural order, the structural constraints imposed between individual pairs of regions may be used to resolve this ambiguity. Irrespective of whether it plays an inhibiting or supportive role, each match can be considered as a source of contextual information about other matches and may therefore aid their interpretation.

Structural constraint may be expressed as the degree to which pairs of matches are mutually compatible. Sources of such constraint are usually derived from world knowledge such as *uniqueness, continuity, topology* and *hierarchy* [8]. The degree of compatibility between any pair of matches γ and γ' is captured by expression

$$-1 \leq C(\gamma, \gamma') \leq 1 \tag{3.1}$$

Pairs of correct matches should ideally enjoy a strong level of mutual compatibility while pairs containing false matches should generate low levels of compatibility.

Structural Matching in Remote Sensing as an Optimisation Problem 17

In our case structural matching employs topological information. A range of topological constraints has been implemented in various stereo systems. Topological constraints attempt to use the fact that the structure of the scene is similar in both images. The relative position of the features is the most widely reported source of constraint. This topological constraint assumes that the vector between features centroid (angle and distance) between images should remain highly similar between pair of true matches.

The typical case of line segments can be used to illustrate the method using an arbitrary example of calculating the degree of compatibility between any pair of matches. The compatibility between two matches may be based on a comparison of the distance and angle between line segments in each image - see figure 3.1.

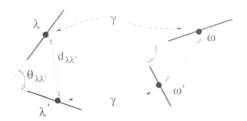

Fig. 3.1. Structural compatibility of line segments.

For compatible matches $\gamma = \{\lambda, \omega\}$ and $\gamma' = \{\lambda', \omega'\}$, the inter-centroid distance and angle between line segments $\lambda \to \lambda'$ and $\omega \to \omega'$ should be similar. The distance d and angle θ between centroids may be calculated as

$$\begin{aligned}
d_{\lambda\lambda'}^2 &= (x_\lambda - x_{\lambda'})^2 + (y_\lambda - y_{\lambda'})^2 \\
\theta_{\lambda\lambda'} &= (\theta_\lambda - \theta_{\lambda'}) \\
d_{\omega\omega'}^2 &= (x_\omega - x_{\omega'})^2 + (y_\omega - y_{\omega'})^2 \\
\theta_{\omega\omega'} &= (\theta_\omega - \theta_{\omega'})
\end{aligned}$$

where (x_λ, y_λ), $(x_{\lambda'}, y_{\lambda'})$, (x_ω, y_ω), $(x_{\omega'}, y_{\omega'})$, θ_λ, $\theta_{\lambda'}$, θ_λ and $\theta_{\lambda'}$ are the midpoints and orientations of the line segments λ, λ', ω and ω' respectively. A match compatibility term may be calculated for the angular and distance disparities

$$C^{(d)}(\gamma, \gamma') = 1 - 2\left(\frac{|d_{\lambda\lambda'} - d_{\omega\omega'}|}{\max(d_{\lambda\lambda'}, d_{\omega\omega'})}\right); \quad -1 \leq C^{(d)}(\gamma, \gamma') \leq 1 \quad (3.2)$$

$$C^{(\theta)}(\gamma, \gamma') = \cos(\theta_{\lambda\lambda'} - \theta_{\omega\omega'}); \quad -1 \leq C^{(\theta)}(\gamma, \gamma') \leq 1 \quad (3.3)$$

and may be combined to create a single source of compatibility information $C(\gamma, \gamma')$ by averaging the angle and distance compatibilities as follows

$$C(\gamma,\gamma') = \begin{cases} -1 & \text{if } \lambda = \lambda' \text{ or } \omega = \omega', \\ \alpha_d C^{(d)}(\gamma,\gamma') + \alpha_\theta C^{(\theta)}(\gamma,\gamma') & \text{else.} \end{cases} \tag{3.4}$$

where α_d and α_θ control the relative influence of each structural component such that $\alpha_d + \alpha_\theta = 1$. Note that the *uniqueness* constraint has also been invoked to ensure a high level of incompatibility between any two matches which share the same feature. A more formal approach based on the joint density functions of the binary measurements has been employed by Christmas *et al.* [2].

4. Matching within an Optimisation Framework

Optimisation has been used extensively in stereopsis. It provides an attractive basis for motivating the selection of the required matches. A cost function may be defined which captures matching information from a wide variety of sources. Comparison of the attributes of the features is a commonly used source which supports correspondence between similar features. Information supporting a candidate match can also be derived from other candidate matches *i.e. compatibility*. In the following section, an optimisation functional is defined while an optimisation method based on *Hopfield Neural Networks* is presented in section 4.

The most common optimisation functional derives directly from a relational graph distance defined by Davis [4]: the distance associated with a mapping M between any two image relational graphs (constructed from attributed features and the attributed binary relations) is defined as the sum of the attribute differences between the attributes of mapped nodes and relations. This is equivalent to maximising a functional whose terms measure and relational consistency of tokens mapped onto each other under the mapping M. Thus for the set of $\mathcal{M}_\tau$ candidate matches the correct matches may be derived from the set of match probabilities

$$\mathbf{p} = \left(p(\gamma_1), p(\gamma_2), ..., p(\gamma_{N_\gamma}) \right) \tag{4.1}$$

which maximises the functional

$$E(\mathbf{p}) = -\alpha_S \sum_{\gamma \in \mathcal{M}_\tau} S(\gamma) + \alpha_C \sum_{\gamma \in \mathcal{M}_\tau} \sum_{\gamma' \in \mathcal{M}_\tau;\ \gamma \neq \gamma'} C(\gamma,\gamma')p(\gamma)p(\gamma') \tag{4.2}$$

where α_S and α_C control the relative influence of the two terms. Having defined an optimisation function, an optimisation technique which will recover the set of probabilities $\mathbf{p}$ which maximises the above expression is required. The iterative neural network algorithm described below starts with the initial set of probabilities $\mathbf{p}_0$ recovered using the attribute comparison method described in section 2.1 (specifically the probabilities computed using equation 2.2).

Structural Matching in Remote Sensing as an Optimisation Problem 19

Updated probabilities must be constrained to lie within the probability solution space. A feasible solution space $\mathcal{P}'$ may be defined as the hypercube of unit width in $\mathcal{P}$ whose vertices represent all possible unambiguous match assignments.

$$0 \leq p(\gamma) \leq 1, \quad \forall \gamma \in \mathcal{M}_\tau \tag{4.3}$$

Ideally, features within one image are required to match uniquely with features in a second. Let a match γ be constructed from the image tokens λ and ω from the first and second image respectively. While $\mathcal{M}_\tau$ denotes the set of all matches, the match sets $\mathcal{M}_\lambda$ and $\mathcal{M}_\omega$ represent the set of competing matches containing the image tokens λ and ω respectively. Then, two further conditions on the optimisation may be defined as

$$\sum_{\gamma \in \mathcal{M}_\lambda} p(\gamma) = 1, \quad \forall \lambda \in \Lambda \tag{4.4}$$

$$\sum_{\gamma \in \mathcal{M}_\omega} p(\gamma) = 1, \quad \forall \omega \in \Omega \tag{4.5}$$

which may be used to define an additional functional term as follows

$$U(\mathbf{p}) = \sum_{\lambda \in \Lambda} \left\{ 1 - \sum_{\gamma \in \mathcal{M}_\lambda} p(\gamma) \right\}^2 + \sum_{\omega \in \Omega} \left\{ 1 - \sum_{\gamma \in \mathcal{M}_\omega} p(\gamma) \right\}^2 \tag{4.6}$$

Incorporating this last pair of constraints is problematic as it is common for tokens within images to be either missing or fragmented. In the case of missing tokens, null matches must be provided if these constraints are to be used. Moreover if a token from one image is required to match two or more subparts in a second image, then this constraint unfortunately makes unambiguous match assignments of the form $p(\gamma) = 0$ or $p(\gamma) = 1$ impossible.

Neural net methods have been successfully applied to an increasing range of visual computing applications including optimisation methods [10, 11, 12]. A neural net consists of a large number of *neurons* where each neuron has an external input and is connected via excitory and inhibitory connections to neighbouring (or all other) neurons. In our problem, each potential match pair may be viewed as such a neuron. A steady state external input potential for each may be derived from the match similarity constraints. Each neuron is connected to other match neurons by an interconnection potential whose strength is related to the inter-match compatibility.

Models describing the behaviour of such a neural net or *network* have been well established [5, 6]. When the network has symmetric interconnection signals, it has been shown to converge to stable output states. It is this stable behaviour (and hence the network's ability to capture *memories*) that lies behind the methods success. Hopfield introduced the following circuit notation to describe the neural network. The external input potential for the ith neuron is given by I_i, and the interconnection *impedances* or weights are denoted by T_{ij}. The network moves to stable state over a finite period of time. The potential at each neuron at time t is given by the equation $u_i(t)$, while the output potential V_i is a function of the neuron potential *i.e.* $V_i = g(u_i)$. This function $g()$ is the *gain* function of the neuron, and is sigmoidal in shape *i.e.* $g(u_i)$ is an increasing monotonic function of u_i. Figure 2.2(b) illustrates a number of such sigmoidal shapes generated by the following typical gain function

$$V = \frac{1}{1 + \exp(-\beta u)} \tag{4.7}$$

where the gain parameter β regulates the speed of convergence.

The general behaviour of each neuron in such a network has been shown by the above authors to be described by the following flow equation

$$C\frac{du_i}{dt} = I_i - \frac{u_i}{R} + \sum_j T_{ij}V_j \tag{4.8}$$

The input *capacitance* C and *resistance* R of the neural inputs determines the convergence rate of the network. Hopfield shows that the network converges to stable states which are local minima in the following function

$$E = -\frac{1}{2}\sum_i\sum_j T_{ij}V_iV_j - \sum V_iI_i \tag{4.9}$$

Note that the form of this function is identical to the cost function of equation 4.2 where the potential $V_i = p(\gamma_i)$, the input current $I_i = S(\gamma_i)$a, and the interconnection impedance T_{ij} is equivalent to the binary terms as defined in equation 3.4. The input external potential I_i is not defined in our case although some suitably normalised similarity term could be used. Consequently

$$\begin{aligned} V_i &= p(\gamma_i) \\ I_i &= -S(\gamma_i) \\ T_{ij} &= -C(\gamma_i, \gamma_j) \end{aligned}$$

Thus the stable state of the network (*i.e.* the resultant node potentials at convergence) is equivalent to the maximum of the match cost function of equation 4.2. The neural net, therefore, appears an ideal method of optimising our cost function subject to the probability *feasibility space*. Note however that the flow function of equation 4.8, requires an estimate of the neuron potential u_i in order to compute du_i/dt. This implies having some initial potential estimate. In most neural implementations however, this question is ignored as some stable state is guaranteed. The network is thought of as storing some *memory*, and that an initial partial memory inputted as the potentials I_i; $i = 1, N$, will be enough to locate the full memory state. Note however that just as the more general optimisation scheme may define

Structural Matching in Remote Sensing as an Optimisation Problem 21

a cost function with more than one local maximum, the network too contains more than one stable states.

The convergence algorithm is implemented as follows. A discrete version of equation 4.8 is computed from the substitutions $du_i/dt = (u_i^{(s+1)} - u_i^{(s)})/\Delta t$ and $\Delta t = 1$ *i.e.*

$$u_i^{(s+1)} = u_i^{(s)} - \frac{1}{C} \sum_j C(\gamma_i, \gamma_j) p(\gamma_j) \qquad (4.10)$$

Assuming that $RC \to \infty$, the step size of this update rule is given by the constant $1/C$. The final neural output potentials $V_i^{(\infty)}$ are computed from the gain function as $g(u_i^{(\infty)})$. The initial estimate of the output potentials $p_0(\gamma_i)$ is known. Thus the initial values $u_i^{(0)}$ may be computed using the inverse gain function $u_i = g^{-1}(p_0(\gamma_i))$.

5. Conclusions

There are three distinctive and independent aspects of matching processes: *token type, sources of constraint* and *method of optimisation*. With this type of approach to reviewing the field it is now possible to assess, compare and, perhaps, experimentally evaluate the contributions made by any particular researcher.

Sources of constraint may be termed *unary, binary* or *N-ary* depending on the number of individual candidate matches they affect. Explicit *N-ary* constraints are rarely implemented. Unary constraints may be derived from the comparison of feature attributes. Specific binary constraints are generally derived from four general world constraints such as *uniqueness, continuity, topology* and *hierarchy*.

In general, the application of unary constraints alone is insufficient to establish a unique set of matches. Binary constraints, therefore, must be employed within some *method of optimisation* scheme to further prune the match set. Such an explicit optimisation framework has the advantage of providing a formal framework within which match constraints may be embedded. Most methods employing optimisation explicitly adopt very similar quadratic functionals which can handle unary and binary data. However a variety of minimisation procedures have been adopted which either iteratively search match probability space from local minima directly (*gradient-descent, relaxation labelling, simulated annealing, neural networks, etc*) or search directly for n-tuples of matches (corners of the probability space hypercube) which minimise the functional.

References

1. B. Chebaro, A. Crouzil, L. Massip-Pailhes, and S. Castan, "A stereoscopic sequence of images managing system", in *Proceedings of the Scandinavian Conference on Image Analysis*, Tromso, Norway, pp. 739–746, 1993.
2. W. J. Christmas, J. Kittler, and M. Petrou, "Modelling compatibility coefficient distributions for probabilistic feature-labelling schemes", in *Proceedings of the British Machine Vision Conference*, Birmingham, pp. 603–612, September 1995.
3. W. J. Christmas, J. Kittler, and M. Petrou, "Structural matching in computer vision using probabilistic relaxation", *IEEE Transactions on Pattern Analysis and Machine Intelligence*, vol. 17, no. 8, pp. 749–764, 1995.
4. L. S. Davis, "Shape matching using relaxation techniques", *IEEE Transactions on Pattern Analysis and Machine Intelligence*, vol. 1, pp. 60–72, January 1979.
5. J. Hopfield and D. Tank, "Neural networks and physical systems with emergent collective computational abilities", *Proceedings of the National Academy of Sciences, USA*, vol. 79, pp. 2554–2558, April 1982.
6. J. Hopfield, "Neurons with graded responses have collective computation properties like those of two-state neurons", *Proceedings of the National Academy of Sciences, USA*, vol. 81, pp. 3088–3092, 1984.
7. R. Horaud and T. Skordas, "Stereo correspondence through feature grouping and maximal cliques", *IEEE Transactions on Pattern Analysis and Machine Intelligence*, vol. 11, pp. 1168–1180, November 1989.
8. G. A. Jones, "Constraint, optimisation and hierarchy: reviewing stereoscopic correspondence of complex features", *Computer Vision and Image Understanding*, vol. 65, pp. 57–78, January 1997.
9. L. E. P. Kierkegaarde, "Stereo matching of curved segments", in *Proceedings of the Scandinavian Conference on Image Analysis*, Tromso, Norway, pp. 457–464, 1993.
10. S. Z. Li, "Relational structure matching", VSSP-TR–4/90, VSSP, Surrey University, Guildford, UK, 1990.
11. R. Mohan and R. Nevatia, "Using perceptual organisation to extract 3D structures", *IEEE Transactions on Pattern Analysis and Machine Intelligence*, vol. 11, pp. 1121–1139, November 1989.
12. N. M. Nasrabadi and C. Y. Choo, "Hopfield network for stereo vision correspondence", *IEEE Transactions on Neural Networks*, vol. 3, pp. 5–13, January 1992.
13. S. A. Pollard, J. Porrill, J. E. W. Mayhew, and J. P. Frisby, "Matching geometrical descriptions in three-space", *Image and Vision Computing*, vol. 5, pp. 73–78, May 1987.
14. R. C. Wilson and E. R. Hancock, "Structural matching by discrete relaxation", *IEEE Transactions on Pattern Analysis and Machine Intelligence*, vol. 19, pp. 634–648, June 1997.
15. M. Xie and M. Thonnat, "A cooperative algorithm for the matching of contour points and contour chains", in *Proceedings 6th International Conference on Image Analysis and Processing*, Como, Italy, pp. 326–333, World Scientific, September 1991.
16. B. Zagrouba, C. J. Krey, and Z. Hamrouni, "Region matching by adjacency propagation in stereovision", in *Proceedings of International Conference on Automation, Robotics and Computer Vision*, Singapore, pp. CV8.5.1–CV8.5.5, The Institution of Engineers, September 1992.

Detail-Preserving Processing of Remote Sensing Images

Silvana Dellepiane

Department of Biophysical and Electronic Engineering (DIBE)
University of Genoa, via Opera Pia 11a, I-16145, Genova, ITALY.
e-mail: silvana@dibe.unige.it

Summary. This paper deals with the problem of detail-preserving processing of remotely sensed images. It describes two approaches based, on the use of the local adaptivity properties exploited by methods of the contextual type (for instance, the Markov Random Field model [6]) and on the application of fuzzy topology by the so-called isocontour method [2], respectively.

1. Introduction

The digital processing of remote sensing images often represents a prerequisite for subsequent interpretation, information extraction, data fusion, change detection, etc. Possible inaccuracies or errors in the restoration. filtering and segmentation phases are therefore very critical for satisfactory results. Indeed, given the new advent of high-resolution remotely sensed images, precise location of structures plays a fundamental role.

From this perspective, the present paper deals with the problem of detail preservation in image processing. This problem is very critical, as a trade-off must be made between preserving fine structures (and small details) and the need for statistically significant results related to neighbourhood dimensions. The use of contextual information requires the analysis of local neighbourhoods, which must be of certain dimensions in order to be statistically significant, but at the same time, they must be small enough to avoid the cancellation of small details.

1.1 The Use of Context in Image Segmentation

The use of context in image processing has proved to contribute appropriately in solving some image analysis problems, such as restoration and segmentation. In texture images, spatial relationships carry all the useful information. In general, even in texture-free images, through spatial continuity one may overcome ambiguities in areas with very low S/N ratios.

Two kinds of contextual information are mainly involved in the processing and segmentation of real images: second-order intensity statistics (which describe spatial arrangements and correlations in the original random field) and class homogeneity (represented by the label configuration in a pixel neighbourhood). The former information is inherent in the original image and

24   S. Dellepiane

does not depend on image processing. It is usually exploited by comparing acquired data with spatial class models. The latter information, on the contrary, is dynamic as label configurations in pixel neighbourhoods represent the non-observable information one aims to obtain as a processing result. Moreover, the former contextual information does not require any prohibitive computational load, even for an exhaustive application of a large number of models. By contrast, exploitation of the latter context represents the bottleneck of the current state of the art, as it results in a combinatorial problem to be solved by a specific method. The need for evaluating label configurations, without imposing any constraint, calls for iterative methods (like those applied by the Markov Random Field (MRF) approach [6]), by which various configurations are assumed, changed, and evaluated. Computation time and convergence are the most critical problems.

As an alternative, in order to solve the computation and convergence problems, the imposition of some constraints on label configurations can be suggested. As a matter of fact, the concept of fuzzy topology provides a solution to contextual processing as it bypasses the iterative processing mechanism.

1.2 Paper Content

Two main approaches are presented in the paper. They exploit two different ideas about the use of context and the preservation of details.The common underlying concept is the adaptivity of spatial analysis to a local image content. A first idea, described in section 2, is based on the local adaptivity properties introduced in image-processing methods of the contextual type, like, for instance, the MRF model.

The proposed ideas represent a change in classical contextual processing techniques, which apply the neighbourhood concept blindly, through the use of a fixed mask shape. The novel concept of adapting the neighbouring mask shape to a local context so as to make the shape change from one image to another, or even inside the same image, is presented for the segmentation of remotely sensed images, and was previously adopted in some published works [13].

The second approach, described in section 3, represents a completely original method, which exploits fuzzy topological features for the detection of very thin structures, in the presence of noise and low contrast. The method allows one to analyze an image along one-dimensional paths at high topological values. Along such paths, the filtering and/or segmentation processes can be performed, without losing details or small structures, even when they are very close to one another.

A more precise location of region boundaries and a better preservation of roads, paths, and other thin structures can be achieved by the two approaches through an analysis of multipolarimetric Synthetic Aperture Radar (SAR) data. A precise processing of interferometric SAR images is also suggested.

In such images, a correct detection of fringes is a prerequisite for the phase-unwrapping stage, and the presence of very close interferometric fringes is very appropriate to prove the applicability of the proposed approach. Along with the descriptions of the two proposed methods, some results obtained with real images are reported.

2. Local Adaptivity in MRF Image Processing

The Gibbs-MRF equivalence allows one to include both types of the context information mentioned above, in a very simple energy term, through two additive energy terms, each related to one context.

The search for the maximum a-posteriori probability (MAP) estimate becomes an optimization problem, whose objective function is exactly the energy or potential term. The solution of such a combinatorial problem is very widely approached through the simulated annealing method, which applies appropriate sampling on label configurations in order to minimize the energy term (i.e., maximize the potential term), while avoiding local maxima.

2.1 Overview of the MRF Model with the Adaptive Neighbourhood Segmentation Approach

In general, the addressed problem is one of finding the true label values for the image lattice S defined on a graph G, where a so-called neighbourhood set N_S is the mapping that associates a neighbourhood with each site. This search for the true label values is accomplished by using an MRF model defined on the neighbourhood set. The proposed method not only minimizes the defined Gibbs function in terms of energy, but it also determines what shape of a neighbourhood set is the most appropriate for the pixel under analysis. Hence we can say that both the shapes of the neighbourhood set and the region label image are optimized during processing. As the iterative minimization of the Gibbs-MRF energy function proceeds, the first thing to be done at each iteration is to determine, for the pixel under analysis and from the information available, the most probable shape of the neighbourhood to be used in the MRF region label process. Then, the selected neighbourhood set is utilized to compute the difference between the energy as a result of the old configuration and the energy of a stochastically proposed new configuration. This difference gives rise to acceptance or rejection, according to the simulated annealing-Metropolis approach. After each iteration, the region label image is updated and converges to a solution, until a stop criterion is reached.

Bayesian inference is used to determine the optimal Markovian neighbourhood set. Although Perez and Heitz [10] provided the proof that one can fully describe the concept of Markovianity by means of clique potentials, we feel that the Bayesian inference approach has advantages in terms of implementation flexibility and possibility of reducing the search space heuristically.

2.2 Multi-Channel Image Statistics and Classical MRF Region Label Model

The maximum a-posteriori probability classifier includes two energy terms. One term describes the statistics of the image data, the other describes the smoothness prior, modelled into an MRF region label process. We consider the measurement vector $X_S = [x_1 \, x_2 \, x_3]_S$, containing for this example the three intensity values of three image channels. By using the Gibbs equivalent of probability, an energy function is defined that expresses the distance of an unknown feature vector X_S to the class model of class l:

$$U_1^S(X_S/L_S = 1) = X_S^T C_l s^{-1} X_S + \ln |C_l s| \tag{2.1}$$

assuming the measurement vector to be circular Gaussian. $C_l = \langle X^* X^T \rangle_l$ is the 3×3 correlation matrix of the class data l. This model does not take into account any spatial information, but allows to define the image class models using the interaction between the different channels.

The second energy term is the so-called smoothness-prior. In the classical case, the conditional distribution of the region label L_S, given the region labels elsewhere, is given by Rignot and Chellappa [11]:

$$U_2^S(L_S = l/L_r, r \in N_S^0) = -\frac{\beta}{N} \sum_{r \in N_S^0} \delta_k(L_S - L_r) \tag{2.2}$$

This is called the region label process, where δ_k is the Kronecker delta function, which returns the value 1 if L_S is equal to L_r and the value 0 otherwise; $N = |N_S^0|$ is the size of the neighbourhood N_S^0 which does not include the current site s. When the attraction parameter β is zero, the image model maximizes the maximum likelihood (ML) probability.

2.3 An MRF Region Label Model with Adaptive Neighbourhoods

The application of a belief function for the Markovian property is founded on a more precise interpretation of the definition of Markovianity, in which the shapes of the neighbourhood sets need not be fixed for the whole image. With this in mind, we could define, for every location in the image, the local Markovian neighbourhood.

Observation 1 (Smits and Dellepiane [13]): in an MRF region label process, defined on a lattice S, a set of pixels $N_S{}^\alpha \subseteq N_S{}^* \subseteq S$ is called a strictly Markovian neighbourhood or a minimum neighbourhood if and only if the set $t \in N_S^\alpha$ for which

$$p(L_S/L_r, r \in N_S^\alpha, r \neq t, L_t) = p(L_S/L_t) \tag{2.3}$$

is empty. In other words, all the elements in the neighbourhood set are meaningful. In the following, N_S^α will be called the Adaptive Neighbourhood (AN) system. To test whether a proposed neighbourhood set is efficient or not, the *hypothesize-and-test* paradigm is used. The hypothesis can be formulated as follows. Let the proposition $A_S = N_S^\alpha$ mean "N_S^α *is the minimum neighbourhood in the Markovian sense*". Such a hypothesis is important, since we are looking for a subset of connected labels. This connectivity between labels is seen as a Markovian relation that does not necessarily extend isotropically over the "standard" neighbourhood.

In the final analysis, the use of a Bayesian belief network is introduced as a substitution for the clique potential functions. The belief value $BEL(N_S^\alpha)$ expresses the belief in the proposition $A_S = N_S^\alpha$ given all the evidence. The evidence can be obtained from various sources of information, as described in [13].

Assumption 1. A connected neighbourhood set $N_S^\alpha \subseteq N_S^*$ is believed to be Markovian if $\forall \epsilon \neq \alpha\ BEL(N_S^\alpha) > BEL(N_S^\epsilon)$.

Assumption 1 states that the proposed shape of a neighbourhood set at the pixel site s is accepted only if its belief value is higher than those of alternative shapes. In the case where none of the alternative hypotheses is accepted, the system uses the null hypothesis of A^0 (figure 2.1, neighbourhood set 0). Given assumption 1, equation 2.2 can be rewritten as:

$$U_2^S(L_S = l/L_r, r \in N_S^\alpha) = -\frac{\beta}{dim(N_S^\alpha)} \sum_{r \in N_S^\alpha} \delta_k(L_S - L_r) \quad (2.4)$$

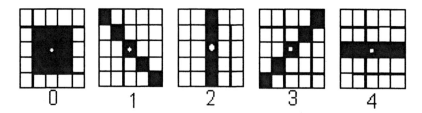

Fig. 2.1. The 5 types of neighbour masks considered in the presented application.

2.4 MAP Segmentation by Simulated Annealing

MAP estimate of L, expressed in terms of energy, minimizes

$$E_{MAP} = \sum_S \{U_1^S(X_S/L_S = 1) + U_2^S(L_S = l/L_r, r \in N_S^\alpha)\} \quad (2.5)$$

For the MAP segmentation, the simulated annealing-Metropolis algorithm is used. Let l be the true label for the site s. Then we may expect that,

$$p(L_S = l/L_r, r \in N_S^\alpha, X_S) \geq p(L_S = l/L_r, r \in N_S^0, X_S), \quad (2.6)$$

28   S. Dellepiane

and therefore that the MAP estimate will converge faster with the adaptive neighbourhood sets. To sum up, the MRF-AN prior is characterized by a more careful application of the smoothness constraint imposed on the label image. It allows the smoothness constraint to be guided by information present in the image or by some extra source of information. Suited support functions have to be defined that guide the decision making process for the neighbourhood set to use.

2.5 Local Adaptivity in MRF Image Processing: Results

The reported results are based on imagery obtained from the Daedalus 1268 Airborne Thematic Mapper (ATM) scanner. It consists of a 200 × 200 pixel portion of the original image and the ATM bands 7, 9 and 10. All data sets contain agricultural areas as well as small man-made objects, such as roads and channels. Six statistical image class models were computed, through the k-means algorithm, based on unsupervised training samples taken from bands 7, 9 and 10, respectively. Figure 2.2a represents the composite of the three bands (in gray-level and enhanced for visualization), figure 2.2b shows the MRF segmentation result, and figure 2.2c the MRF-AN segmentation result. In figure 2.2d it is shown that the selection of the non-standard neighbourhoods takes place at region borders and along fine structures.

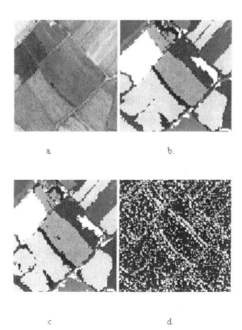

Fig. 2.2. a) Original data set, b) K-means based unsupervised MRF segmentation result, c) Like b) but with MRF-AN, d) Image of the neighbourhood activities.

In conclusion, we can state that the MRF-AN is a modification of the segmentation method proposed by Rignot and Chellappa [11]. The modification concerns the integration of various sources of information to choose the shape of a neighbourhood that is the most appropriate for the local image around the pixel site under test. The choice is based on a belief value of the proposition that the proposed neighbourhood is a minimum Markovian neighbourhood. From among the different shape alternatives, the one with the highest belief value is chosen to compute the MRF region label process. In other papers (see for instance [14]), the concept of the MRF-AN region label model has been compared with the concept of the Discontinuity Adaptive MRF model (DA-MRF), proposed in [7]. Comparing the two concepts it has been shown that the MRF-AN prior model can be regarded as a special instance of the DA-MRF prior.

3. Fuzzy Topology for Image Segmentation: Isocontours

Despite the difficulties in implementing fuzzy methodologies for image processing at the numerical level, the use of fuzzy-set theory seems very appropriate to cope with ambiguities and uncertainties in the image model, particularly in real domains such as remote sensing, where objects are very often ill-defined, due to the physical characteristics of imaging systems and to small differences in energy absorption by different types of landcovers or tissues. Therefore, an exact location of contour points is almost impossible and thin details may be lost.

In the proposed approach, fuzzy topological measures are exploited, thus providing a fuzzy label value to each element of a multichannel image data set. By using the proposed segmentation method, one can perform an analysis of an image signal, without any specific hypothesis about the underlying image or physical model. However, one can solve the problem of preserving details during segmentation, by applying a fuzzy approach to uncertainty handling and by exploiting contextual information through an analysis of spatial configuration relationships.

The method is based on the newly defined concept of fuzzy-intensity connectedness (also named χ-connectedness), fully described in [2], and on a new image scanning mechanism that has been appropriately designed to extract a connectivity measure for each image element, thus avoiding an iterative procedure. The generalization of the basic concepts of crisp geometry to the theory of fuzzy sets was suggested by Rosenfeld [12], who fuzzified a set of tools for describing the geometries of fuzzy subsets, i.e. images and regions in a fuzzy space. Among others, the concept of fuzzy connectivity was defined, that seemed to open new vistas for the description of region topology. Since then, very few papers have appeared in the literature about the use of fuzzy geometry for image-processing tasks ([8, 9]): however, these papers did not deal with connectivity. We believe that this has mainly been due to the

30   S. Dellepiane

following problems: i) the limited generality of the definition of fuzzy connectivity (i.e., the definition nicely applies to tops and bottoms of fuzzy subsets but it does not consider intermediate grey levels); ii) the intrinsic difficulty with computing fuzzy connectivity. The computation of fuzzy connectivity is based on the detection of the best path between two arbitrary points, and this task is clearly very complex to accomplish. The proposed approach allows us to exploit the potentialities of fuzzy intensity connectivity and to generalize them to prove the capabilities for preserving small details. Let us describe a generic fuzzy field $M = \{\mu(p)\}$ over a characteristic function $\mu(p)$, which assigns a real value ranging in the interval $[0, 1]$ to each element p in the fuzzy space If we define the path $P(q, p)$ as a connected sequence of points from the pixel q to the pixel p, the conventional fuzzy degree of connectedness from q to p is expressed as follows (see Rosenfeld [12]):

$$c_\mu(q, p) = c(\mu, q, p) = \max_{P(q,p)} [\min_{z \in P(q,p)} \mu(z)] \qquad (3.1)$$

where the *max* is applied to all the paths $P(q, p)$ from q to p (i.e., the best paths connecting p to q), and the *min* is applied to all the points z along the path $P(q, p)$. This definition implies that connectivity is symmetrical, transitive and weakly reflexive. The definition of degree of connectedness refers to the fuzzy field M. The same definition can be applied directly to the fuzzy field $H = \{\eta(p)\}$ derived from the original image field, as every digital picture can be represented by an equivalent fuzzy field. As stated in [12], there always exists a rational value r such that, for each pixel p, one has:

$$\eta(p) = r \cdot \zeta(p) \qquad (3.2)$$

According to this property, application of the connectedness formula to the field H leads to derive the degree of connectedness from the intensity levels. In other words, a fuzzy-connected component can be identified for each image point, in terms of grey-level homogeneity. However, the definition given in equation 3.1 suffers from lack of generality, as it implies that it is possible to derive connectivity paths only from the tops in an image or from the bottoms in the complement image. To consider also middle-ranged values, the classical fuzzy definition is applied here to a modified field. The new field with respect to the seed point a (i.e., $X^a = \{\chi^a(p)\}$) is expressed as:

$$\chi^a(p) = 1 - |\eta(p) - \eta(a)| \qquad (3.3)$$

where $\eta(p)$ denotes the fuzzy field equivalent to the original image. This leads to compute:

$$c_{\chi a}(p) = c(\chi^a, a, p) = \max_{P(a,p)} [\min_{z \in P(a,p)} \chi^a(z)] \qquad (3.4)$$

which we call intensity connectedness. $P(a, p)$ denotes a path from a to p. The above formula corresponds to the traditional formula for a seed placed at a top where $\eta(a)$ is equal to 1, and for a seed placed at a bottom (at level 0)

Detail-Preserving Processing of Remote Sensing Images 31

in the complement image. As an important property, it is interesting to note that intensity connectedness is a measure consistent with Measure Theory as, when applied to the field X^a, it is invariant to any change in the η axis.

3.1 Selective Image Scanning

The selective scanning mechanism proposed and fully described in [2] is able to propagate local and global contexts via a single iteration by following maximum-connectedness paths. This procedure generates a tree, whose root is the seed point. The root is subsequently expanded so as to process first the pixels at high degrees of membership. In such a way, each image pixel is grown according to an adaptive order. As a side effect of the performed processing, each tree path from the root to a leaf corresponds to the best path from the seed (in terms of intensity connectedness). In other words, the intensity connectedness at the site i is equal to its value at χ^a, or to the intensity connectedness value of its parent, if its intensity connectedness has not the minimum value along the best path from a to i. A direct consequence of this situation is that the label value shows a monotonic behaviour along each path departing from the seed point.

A multivalued segmentation result can be achieved for each seed (hence for each object). Each point in Γ^a that is characterized (in the field M^a) by a membership value larger than a fixed value α is regarded as belonging to the isovolume at the level α of Γ^a. In accordance with the theory of fuzzy sets, the isoregion at the level α is then the α-level set (or α-cut) $IsoReg_\alpha$:

$$IsoReg_\alpha = \{p|M^a(p) \geq \alpha\} = \{p|\chi^a(p) \geq \alpha\} \tag{3.5}$$

It is ensured that, for any value α and for each seed point a, the isoregions obtained as α-cuts of the connectivity map are always connected regions, thus representing a single structure of interest in the image. An example of isoregion, as extracted from a SAR interferometric image, is given in figure 3.1. The isocontours are obviously derived from the isoregions.

3.2 A Fuzzy-Contextual Segmentation Method

On the basis of the above-described fuzzy isocontours, a fuzzy contextual segmentation (FCS) method has been developed, as reported in [3]. As in the MRF approach, two terms are used, one referring to global region properties, the other representing local continuity. The label is the weighted sum:

$$\mu = \beta\mu_1 + \gamma\mu_2 \tag{3.6}$$

of the terms:

$$\mu_1 = sim(SR) \tag{3.7}$$

and

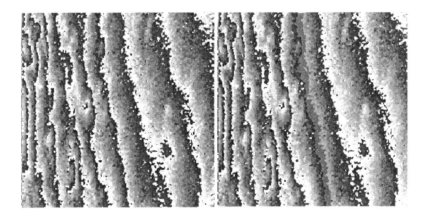

Fig. 3.1. Example of interferometric SAR image and of an isoregion depicting an interferometric fringe.

$$\mu_2 = sim(NN) \cdot \mu_n N \quad (3.8)$$

where $sim()$ denotes a similarity measure with respect to the current site, SR the reference (seed) region, and NN the Nearest Neighbour. The generalization of the isocontour method to a multiseed application can be easily described by introducing the matrix of the prototypes, the global similarity matrix, and the label (or membership) matrix. The c by c prototype matrix is an identity matrix, as each selected reference point is assigned with a hard label to only one class, and used as a prototype. The prototypes will be used as reference constraints in the subsequent processing. The similarity matrix (of size c by n, n being the number of pixel sites) is a fuzzy partition matrix as defined in [1]. It is built by assigning to each image site j its similarity value μ_1^{ij} with respect to all the prototypes i. Since it refers to global references, it is built sequentially and univocally determined as it does not depend on the order analysis. The last matrix, i.e. the label matrix (of the fuzzy type), represents the final processing result. It contains the label value derived from the two terms in equations 3.3 and 3.4 obtained from the multiseed set. The use of a local context is very critical as it depends on the selection of the related global context and on the label of the context itself.

Two main points have been addressed in the design of the method: the criterion for the selection of the Nearest-Neighbour site and the order of analysis (i.e., the image sampling or scanning mechanism). The final decision is made on the basis of the label matrix. Additional features are thus available, as compared with traditional non-fuzzy partition matrices. The entropy value of each column of the matrix represents the ambiguity of the label assigned to the related pixel site. This additional feature can be used to give a degree of certainty to the assignment, (or to backtrack the processing in those areas for which a right decision is more difficult), on the basis of more complete information resulting from preliminary partial results (the ones with smaller

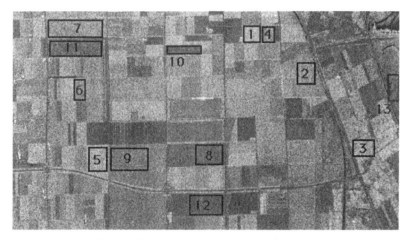

Fig. 3.2. Flevoland Image. Original L-band VV polarization and the thirteen training areas considered for the experiment. The training areas correspond to: 1) potatoes; 2) stem beans; 3) forest; 4) red beets; 5) peas; 6) beets; 7) bare soil; 8) lucerne; 9) winter wheat; 10) grass; 11) flax; 12) summer barley; 13) water.

entropy values). As a final general remark, the assignment formula represents a combination of the k-NP (Nearest Prototype) and k-NN (Nearest Neighbour) paradigms [4], and depends on the values of the parameters β and γ.

3.3 A Fuzzy Contextual Segmentation Method: Results

A 4-look SAR data on the Flevoland area (The Netherlands), acquired from the Maestro 1 campaign, is utilized to assess the performance of the Fuzzy Contextual Segmentation (FCS) algorithms. Figure 3.2 shows the acquired VV polarization in the L-band frequency, together with the training areas considered. The results obtained by the FCS over the HH polarizations channel are given in figure 3.3(a). Figure 3.3(b) shows the results obtained by applying the FCS methods over both the HH and the VV polarization channels. The classification accuracies achieved are reported in figure 3.4. The results of the FCS method were also compared to the results obtained, on the same data set, by the Maximum Likelihood (ML) method (applied only to the HH and VV channels), and by the method described in [5].

4. Conclusions

In this paper we presented two approaches for the solution of some typical image analysis problems, such as segmentation and restoration, through the exploitation of two kinds of contextual information: second order intensity

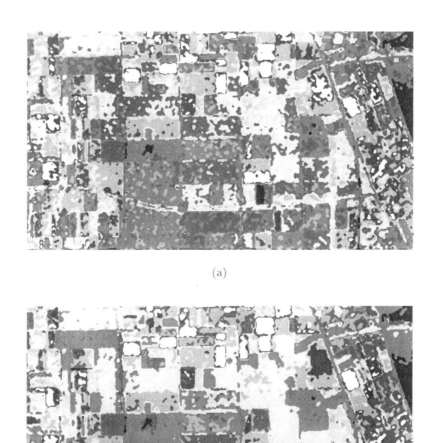

Fig. 3.3. Flevoland Image. Supervised Fuzzy Contextual Segmentation results obtained by using the HH polarization (a), and a combination of the HH and VV polarizations (b).

	ML HH	ML VV	FCS HH	FCS VV	FCS HH+VV	Freeman et al
1. Potatoes	75.5	77.7	89.0	91.7	97.1	92
2. Stem Beans	16.2	43.0	37.3	30.0	91.4	93
3. Forest	11.5	7.3	28.3	20.2	52.5	95
4. Red Beets	42.7	21.6	84.1	70.5	92.7	91
5. Peas	12.4	24.6	16.3	37.0	92.5	96
6. Beets	19.5	45.7	32.0	72.9	77.4	90
7. Bare Soil	32.8	7.8	77.9	16.7	96.0	99
8. Lucerne	48.3	59.8	76.5	84.6	95.2	99
9. Winter Wheat	42.9	30.2	64.9	56.9	100.0	48
10. Grass	14.1	16.7	34.8	26.0	72.7	56
11. Flax	45.9	13.7	31.7	0.0	96.4	99
12. Summer Barley	43.0	58.3	79.5	84.9	98.5	40
13. Water	92.0	25.0	100.0	30.6	93.3	98
Overall	47.0	29.8	57.6	44.3	90.4	

Fig. 3.4. Flevoland Image. Percent classification accuracy. ML and FCS classification accuracies are given for the HH and the VV (ML and FCS) and for the HH+VV polarization (FCS), as compared with the training areas. The last column refers to the accuracy obtained by Freeman [5], on the fully polarimetric, complex data, as compared with the ground truth.

statistics and class homogeneity. The former feature describes the spatial distribution and correlation present in the original data and results to be independent from the actual image processing algorithm. The latter is a dynamic information representing the label configuration in a pixel neighbourhood, and could be obtained by executing an appropriate processing. Unfortunately, the preservation of small details becomes very critical when using contextual knowledge: large neighbourhoods ensure statistical reliability but they cause the loss of thin structures. In order to overcome this problem, we introduced the concept of adaptivity of spatial analysis to the local image content. In the first method an adaptive neighbouring mask has been used for tracking the local data behaviour. The second approach exploits fuzzy topological features for the detection and preservation of thin details. The results obtained by applying both methods to real remote sensing images show the effectiveness of the described approaches.

References

1. J. C. Bezdek, *Pattern recognition with fuzzy objective function algorithms*, Plenum Press, New York, 1981.

2. S. Dellepiane, F. Fontana, and G. Vernazza, "Non-linear image labelling for multivalued segmentation", *IEEE Transactions on Image Processing*, vol. 5, no. 3, pp. 429–446, 1996.
3. S. Dellepiane and F. Fontana, "Supervised fuzzy contextual segmentation of polarimetric SAR images", *European Transactions on Telecommunications*, vol. 6, pp. 515–525, 1996.
4. R. O. Duda and P. E. Hart, *Pattern classification and Scene Analysis*, Wiley Interscience, New York, 1974.
5. A. Freeman, J. Villasenor, J. D. Klein, P. Hoogeboom and J. Groot, "On the use of multi-frequency and polarimetric radar backscatter features for classification of agricultural crops", *International Journal of Remote Sensing*, vol. 15, no. 9, pp. 1799–1812, 1994.
6. S. Geman and D. Geman, "Stochastic relaxation, Gibbs distributions, and the Bayesian restoration of images, *IEEE Transactions on Pattern Analysis and Machine Intelligence*, vol. 6, no. 6, pp. 721–741, 1984.
7. S. Z. Li, "On discontinuity-adaptive smoothness priors in computer vision", *IEEE Transactions on Pattern Analysis and Machine Intelligence*, vol. 17, no. 6, pp. 576-586, 1995.
8. S. K. Pal and A. Rosenfeld, "Image enhancement and thresholding by optimization of fuzzy compactness", *Pattern Recognition Letters*, vol. 8, pp. 21–28, 1988.
9. S. K. Pal and A. Ghosh, "Index of area coverage of fuzzy image subsets and object extraction", *Pattern Recognition Letters*, vol. 11, pp. 831–841, 1990.
10. P. Perez and F. Heitz, "Restriction of a Markov random field on a graph and multiresolution statistical image modelling", *IEEE Transactions on Information Theory*, vol. 42, no. 1, pp. 180–190, 1996.
11. E. Rignot and R. Chellappa, "Segmentation of polarimetric synthetic aperture radar data", *IEEE Transactions on Image Processing*, vol. 1, no. 3, pp. 281–300, 1992.
12. A. Rosenfeld, "The fuzzy geometry of image subset", *Pattern Recognition Letters*, vol. 2, pp. 311–317, 1984.
13. P. C. Smits and S. G. Dellepiane, "Synthetic Aperture Radar image segmentation by a detail preserving Markov Random Field approach, *IEEE Transactions on Geoscience and Remote Sensing*, vol. 35, no. 4, pp. 844–857, 1997.
14. P. C. Smits and S. Dellepiane, "Irregular MRF region label model for multichannel image segmentation", *Pattern Recognition Letters* vol. 18, no. 11–13, pp. 1133–1142, 1997.

Multi-Channel Remote Sensing Data and Orthogonal Transformations for Change Detection

Allan Aasbjerg Nielsen

Department of Mathematical Modelling, Technical University of Denmark, Building 321, DK-2800 Lyngby, Denmark[**].
e-mail: aa@imm.dtu.dk

Summary. This paper describes the multivariate alteration detection (MAD) transformation which is based on the established canonical correlation analysis. It also proposes post-processing of the change detected by the MAD variates by means of maximum autocorrelation factor (MAF) analysis. As opposed to most other multivariate change detection schemes the MAD and the combined MAF/MAD transformations are invariant to affine transformations of the originally measured variables. Therefore, they are insensitive to, for example, differences in gain and off-set settings in a measuring device, and to the application of radiometric and atmospheric correction schemes that are linear or affine in the gray numbers of each image band. Other multivariate change detection schemes described are principal component type analysis of simple difference images. A case study with Landsat TM data using simple linear stretching and masking of the change images shows the usefulness of the new MAD and MAF/MAD change detection schemes. A simple simulation of a no-change situation shows the power of the MAD and MAF/MAD transformations.

1. Introduction

When analysing changes in panchromatic images taken at different points in time, it is customary to examine the difference between the two images after normalisation. The idea is that areas with no or little change have zero or low absolute values and areas with large changes have large absolute values in the difference image. If we have two multivariate images with variables at a given location written as vectors (without loss of generality we assume that $E\{\boldsymbol{X}\} = E\{\boldsymbol{Y}\} = \boldsymbol{0}$)

$$\boldsymbol{X} = [X_1 \ \ldots \ X_k]^T \quad \text{and} \quad \boldsymbol{Y} = [Y_1 \ \ldots \ Y_k]^T \tag{1.1}$$

where k is the number of spectral bands, then a simple change detection transformation is

$$\boldsymbol{X} - \boldsymbol{Y} = [X_1 - Y_1 \ \ldots \ X_k - Y_k]^T \tag{1.2}$$

If our image data have more than three channels it is difficult to visualise change in all of them simultaneously. To overcome this problem, and to concentrate information on change, linear transformations of the image data that

---

[**] http://www.imm.dtu.dk/~aa

optimise some design criterion can be considered. A linear transformation that will maximise a measure of change in the simple multispectral difference image is one that maximises deviations from no change, for instance the variance

$$\text{Var}\{v_1(X_1 - Y_1) + \cdots + v_k(X_k - Y_k)\} = \text{Var}\{v^T(X - Y)\} \qquad (1.3)$$

Areas in the image data with high absolute values of $v^T(X-Y)$ are maximum change areas. A multiplication of vector v with a constant c will multiply the variance with c^2. Therefore we must make a choice concerning v. A natural choice is to request that v is a unit vector, $v^T v = 1$. This amounts to finding principal components of the simple difference images. Other change detection schemes based on simple difference images include factor analysis and maximum autocorrelation factor (MAF) analysis [17].

A more parameter rich measure of change that allows different coefficients for X and Y and different numbers of spectral bands in the two sets, p and q respectively ($p \le q$), are linear combinations

$$a^T X = a_1 X_1 + \cdots + a_p X_p \quad \text{and} \quad b^T Y = b_1 Y_1 + \cdots + b_q Y_q \qquad (1.4)$$

and the difference between them $a^T X - b^T Y$. This measure also accounts for situations where the spectral bands are not the same but cover different spectral regions, for instance if one set of data comes from the Landsat Multi-Spectral Scanner (MSS) and the other set comes from the Landsat Thematic Mapper (TM) or from the SPOT High Resolution Visible (HRV) sensor. This may be valuable in historical change analysis. In this case one must be more cautious when interpreting the multivariate difference as multivariate change.

To find a and b principal components (PC) analysis on X and Y considered as one concatenated vector variable has been used [4]. In [5], PC analysis is applied to simple difference images as described above. The approach in [4] defines a and b simultaneously but the method does not have a clear design criterion. Also, bands are treated similarly whether or not they come from different points in time. The approach in [5] depends on the scale at which the individual variables are measured (for instance it depends on gain settings of a measuring device). Also, it forces the two sets of variables to have the same coefficients (with opposite sign), and it does not allow for the case where the two sets of images have different numbers of channels. A potentially better approach is to define a set of a and b simultaneously in the fashion described below. Again, let us maximise the variance, this time $\text{Var}\{a^T X - b^T Y\}$. A multiplication of a and b with a constant c will multiply the variance with c^2. Therefore we must make choices concerning a and b, and natural choices in this case are requesting unit variance of $a^T X$ and $b^T Y$. The criterion then is: maximise $\text{Var}\{a^T X - b^T Y\}$ with $\text{Var}\{a^T X\} = \text{Var}\{b^T Y\} = 1$. With this choice we have

$$\begin{aligned} \text{Var}\{a^T X - b^T Y\} &= \text{Var}\{a^T X\} + \text{Var}\{b^T Y\} - 2\text{Cov}\{a^T X, b^T Y\} \quad (1.5) \\ &= 2(1 - \text{Corr}\{a^T X, b^T Y\}) \end{aligned}$$

We shall request that $a^T X$ and $b^T Y$ are positively correlated. Therefore, determining the difference between linear combinations with maximum variance corresponds to determining linear combinations with minimum (nonnegative) correlation. Determination of linear combinations with extreme correlations brings the theory of canonical correlation analysis to mind.

Canonical correlation analysis investigates the relationship between two groups of variables. It finds two sets of linear combinations of the original variables, one for each group. The first two linear combinations are the ones with the largest correlation. This correlation is called the first canonical correlation and the two linear combinations are called the first canonical variates. The second two linear combinations are the ones with the largest correlation subject to the condition that they are orthogonal to the first canonical variates. This correlation is called the second canonical correlation and the two linear combinations are called the second canonical variates. Higher order canonical correlations and canonical variates are defined similarly. The technique was first described in [8] and a treatment is given in most textbooks on multivariate statistics (good references are [1, 2]).

Multivariate change detection techniques are also described in [6, 7] both dealing with multiple regression and canonical correlation methods applied to specific change detection. Wiemker *et al.* [18] deal with (iterated) PC analysis of the same variable at the two points in time and consider the second PC as a (marginal) change detector for that variable. The same authors [18] also introduce spatial measures such as inverse local variance weighting in statistics calculation and Markov random field modelling of the probability of change (vs. no change).

2. The Multivariate Alteration Detection (MAD) Transformation

In accordance with the above we define the multivariate alteration detection (MAD) transformation as

$$\begin{bmatrix} X \\ Y \end{bmatrix} \rightarrow \begin{bmatrix} a_p^T X - b_p^T Y \\ \vdots \\ a_1^T X - b_1^T Y \end{bmatrix}, \tag{2.1}$$

where a_i and b_i are the defining coefficients from a standard canonical correlations analysis. X and Y are vectors with $\mathrm{E}\{X\} = \mathrm{E}\{Y\} = 0$. The dispersion matrix of the MAD variates is

$$\mathrm{D}\{a^T X - b^T Y\} = 2(I - R) \tag{2.2}$$

where I is the $p \times p$ unit matrix and R is a $p \times p$ matrix containing the sorted canonical correlations on the diagonal and zeros off the diagonal.

40 A. A. Nielsen

The MAD transformation has the very important property that if we consider linear combinations of two sets of p respectively q $(p \leq q)$ variables that are positively correlated then the p'th difference shows maximum variance among such variables. The $(p - j)$'th difference shows maximum variance subject to the constraint that this difference is uncorrelated with the previous j ones. In this way we may sequentially extract uncorrelated difference images where each new image shows maximum difference (change) under the constraint of being uncorrelated with the previous ones. If $p < q$ then the projection of Y on the eigenvectors corresponding to the eigenvalues 0 will be independent of X. That part may of course be considered the extreme case of multivariate change detection. As opposed to the principal components, the MAD variates are invariant to linear scaling, which means that they are not sensitive to -for example- gain settings of a measuring device, and to linear radiometric and atmospheric correction schemes. This type of multivariate change detection technique was first outlined in [3], see also [10, 13].

To find maximum change areas with high spatial autocorrelation a MAF post-processing of the MAD variates is suggested. The MAF transformation can be considered as a spatial extension of PC analysis in which the new variates maximise autocorrelation between neighbouring pixels rather than variance (as with PCs). Also the MAF transformation is invariant to linear scaling.

For other descriptions of the transformation described here see [9] which describes the MAD transformation and introduces the MAF post-processing of the MADs, [11] which describes the MAD transformation in more detail, and [12] where one case study very successfully applies the MAD transformation on pre-processed AVHRR data (correction for atmospheric water vapour attenuation, and cloud- and land-masking) in an oceanographic case study on ENSO-related (El Niño/Southern Oscillation) mid-latitude warming events where principal components analysis of simple difference images fail. Another case in [12] deals with urbanisation in Australia using Landsat MSS data.

3. Case Study - Landsat TM Data

The applicability of the MAD transformation with the MAF post-processing to multivariate and bi-temporal change detection studies is demonstrated in a case using Landsat-5 Thematic Mapper (TM) data covering a small forested area approximately 20 kilometres north of Umeå in northern Sweden. The data consist of six spectral bands with 512×512 – 20 metre pixels, rectified to the Swedish national grid by the Swedish Space Corporation who also provided the data. The acquisition dates are 6 June 1986 and 27 June 1988. The intention of this case study is to illustrate the method and not to accurately assess the actual change on the ground. These data are also dealt with in [14, 15, 16].

Orthogonal Transformations for Change Detection   41

The images in Figures 3.1 and 3.2 are stretched linearly from mean minus three standard deviations to mean plus three standard deviations. Figure 3.1 shows the original Landsat TM data from 1986 in column one, the original Landsat TM data from 1988 in column two, the simple difference images ("1988 minus 1986") in column three, and the principle components of the simple difference images in column four. Row one is TM1/component one, row two is TM2/component two, etc. The change information from all bands is concentrated in the first three or four PCs. Because of the stretching applied, areas that are very bright or very dark are maximum change areas.

Figure 3.2 shows the canonical variates of the 1986 data in column one, the canonical variates of the 1988 data in column two, the MADs of the original TM images (the differences between the CVs in reverse order) in column three, and the MAF/MADs of the original TM images in column four. Row one is transformed variable one, row two is transformed variable two, etc. Again, areas that are very bright or very dark are maximum change areas. The low order MADs are quite noisy. It is to be expected that scanner noise and differences in atmospheric conditions are among the most different characteristics of the two scenes. In the MAF/MADs, areas that are very bright or very dark are areas of change with high spatial autocorrelation. The change information from all bands is concentrated in the first three or four MAF/MADs.

Figure 3.3 shows negated absolute values stretched from mean (before taking absolute values) plus two standard deviations (before taking absolute values) to mean plus three standard deviations. Column one shows simple differences, column two principal components of simple differences, column three MADs, and column four MAF/MADs. Row one is transformed variable one, row two is transformed variable two, etc. This time bright regions are no-change areas and dark regions are maximum change areas.

In spite of their high information content RGB plots of the first three PCs of simple differences and the first three MAF/MADs are omitted for reproduction purposes.

For all change detection schemes an interpretation of the resulting change images can be based on visual inspection and correlations between the transformed variates and the original data. Table 3.1 shows the correlations between the MAF/MADs and the original Landsat TM bands calculated only where the standardised score of any MAF/MAD variate is outside the plus/minus three standard deviations interval. MAF/MAD1 is positively correlated with 1986 TM4 and uncorrelated with 1986 TM3. MAF/MAD1 is negatively correlated with 1988 TM4 and positively correlated with 1988 TM3. Therefore MAF/MAD1 is a sort of vegetation index change detector. MAF/MAD2 is a change detector of the weighted mean of all bands except TM4, presumably a non-vegetation related change detector. MAF/MAD3 seems to be a non-vegetation related change detector also. MAF/MAD5 and 6 have their highest correlations with TM1 and TM2 and are therefore likely to

Fig. 3.1. 1986 TM data (first column), 1988 TM data (second column), simple differences "1988–1986" (third column), and PCs of simple differences (fourth column).

Orthogonal Transformations for Change Detection 43

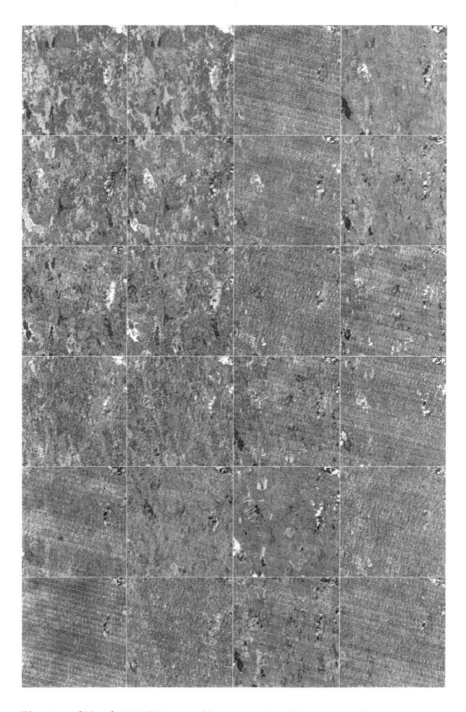

Fig. 3.2. CVs of 1986 TM data (first column), CVs of 1988 TM data (second column), MADs (third column), and MAF/MADs (fourth column).

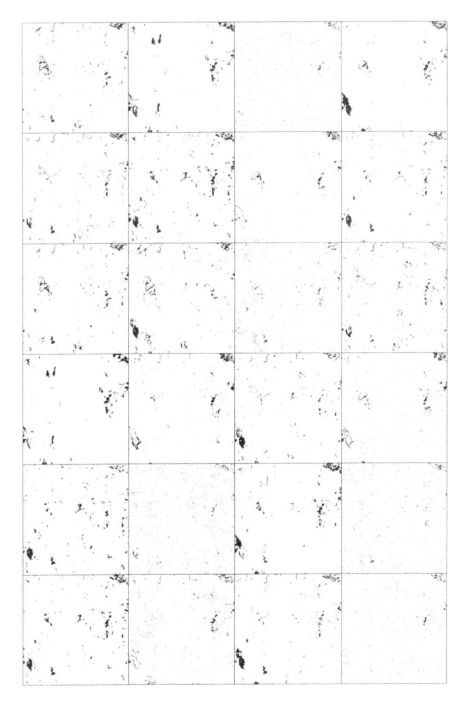

Fig. 3.3. Negated absolute values of simple differences (first column), PCs of simple differences (second column), MADs (third column), and MAF/MADs (fourth column); white represents no-change areas.

Orthogonal Transformations for Change Detection   45

Table 3.1. Correlations between MAF/MADs and Landsat TM bands.

	MAF/ MAD1	MAF/ MAD2	MAF/ MAD3	MAF/ MAD4	MAF/ MAD5	MAF/ MAD6
1986 TM1	0.0089	0.5935	-0.1123	0.0142	-0.0399	0.1668
1986 TM2	0.1818	0.5589	-0.0567	0.1529	0.0260	0.0356
1986 TM3	0.0069	0.6548	-0.1052	0.0469	-0.1466	-0.0197
1986 TM4	0.4557	0.3039	0.0719	0.1630	-0.0217	0.1591
1986 TM5	0.1503	0.6145	0.1020	0.0729	-0.0636	0.0362
1986 TM7	-0.0320	0.6562	0.0816	-0.0990	-0.1077	-0.1077
1988 TM1	0.5343	-0.2053	0.2689	0.0991	-0.2147	-0.0293
1988 TM2	0.3632	-0.0964	0.3141	0.0098	-0.1771	0.1695
1988 TM3	0.6167	-0.2964	0.2591	0.0685	0.0021	0.0319
1988 TM4	-0.4385	0.3866	0.1113	0.0951	-0.0136	0.1285
1988 TM5	0.3976	-0.3634	-0.1730	-0.0676	-0.0137	-0.0148
1988 TM7	0.5884	-0.4006	-0.0838	0.1065	-0.0392	0.0488

represent change in atmospheric conditions. The MAF analysis of the MADs seems to have concentrated the change related to atmospheric conditions and change related to vegetation in each end of the autocorrelation "spectrum". Also, scanner noise is concentrated in the high order MAF/MADs.

3.1 Simple No Change Simulation

As a simple simulation of a situation with no change in all bands we pad the two 512×512 scenes described above into the central part of 600×600 backgrounds with values 0 in all bands in both years. Change between the two 600×600 scenes is estimated by means of 1) simple differences, 2) principal components of simple differences (based on the covariance matrix), 3) principal components of simple differences (based on the correlation matrix), 4) varimax rotated factors of simple differences, 5) MAFs of simple differences, 6) MADs, and 7) MAF/MADs. Change detected in the region with no change (the 44 pixels broad edge around the actual image data) as indicated by standardised values of the results from the above change detection methods is given in table 3.2.

Table 3.2. Change detected (standardised scores) in no change region, value should be 0.

Band/Component	1	2	3	4	5	7/6
1) Simple diff.	0	0	0	0	0	0
2) PC diff. (cov)	-1.28	0.06	0.09	0.91	-0.03	0.26
3) PC diff. (corr)	-0.06	-1.53	0.19	0.40	0.03	-0.01
4) Factors diff.	-0.10	-0.81	1.22	0.22	-0.58	-0.05
5) MAF diff.	-1.52	0.05	0.46	0.03	0.01	0.01
6) MAD	0.01	0.01	0.01	0.01	0.01	-0.03
7) MAF/MAD	-0.00	0.01	0.02	0.00	-0.02	0.02

46   A. A. Nielsen

The value in this region in the simple difference images is zero. Therefore the change detected here by most multivariate methods is due to the subtraction of the mean value of the entire image before calculating the relevant linear combinations of the original bands. This can easily be identified in this very simple situation, but had the no-change pixels been scattered as several fields inside the image this would not have been possible and the statistics for calculating the MADs would have been exactly the same. Therefore the situation simulated, though very simple, is indeed a realistic one.

It is obvious that MAD and MAF/MAD are the only multivariate techniques that perform well in this situation. All other methods give change values much higher than zero.

4. Conclusions

A new approach to change detection in multivariate, bi-temporal, spatial data based on canonical correlation analysis called the multivariate alteration detection (MAD) transformation and maximum autocorrelation factor (MAF) post-processing is described. The success of this scheme is demonstrated in a case with Landsat TM data from a forested region in Sweden. The difference between this scheme and the one based on principal components analysis of the simple difference images suggested in [5] is the invariance of the MAD and MAF transformations to linear and affine transformations which makes the use of the MAD and the combined MAF/MAD transformation insensitive to for instance 1) differences in gain and off-set in a measuring device, and 2) linear radiometric and atmospheric calibration methods. A simple but realistic simulation of a no-change situation favours the MAD and the combined MAF/MAD methods.

Acknowledgement. I wish to thank Professor Knut Conradsen, IMM, for his cooperation on the development of the MAD concept. The calculations were performed by means of software written with Dr. Rasmus Larsen, Novo Nordisk A/S (formerly IMM). Also, I wish to thank Dr. James J. Simpson, Scripps Institution of Oceanography, La Jolla, California, for his immediate interest and comments on the MAD and the MAF/MAD schemes. The Landsat TM data were georectified and kindly provided by the Swedish Space Corporation. Funding from the European Commission and Danida, the Danish international development agency, is highly appreciated.

References

1. T. W. Anderson, *An Introduction to Multivariate Statistical Analysis*, 2nd Edition. John Wiley & Sons, New York, 1984.

2. W. W. Cooley and P. R. Lohnes, *Multivariate Data Analysis*, John Wiley & Sons, New York, 1971.

3. K. Conradsen and A. A. Nielsen, "Multivariate change detection in multispectral, multitemporal images", *Abstracts and Notes from Seminar on Near Real-Time Remote Sensing for Land and Ocean Applications*, Eurimage and ESA/Earthnet, Rome, Italy, 1991.

4. T. Fung and E. LeDrew, "Application of principal components analysis to change detection", *Photogrammetric Engineering and Remote Sensing*, vol. 53, no. 12, pp. 1649–1658, 1987.

5. P. Gong, "Change detection using principal component analysis and fuzzy set theory", *Canadian Journal of Remote Sensing*, vol. 19, no. 1, pp. 22–29, 1993.

6. H. Hanaizumi, S. Chino and S. Fujimura, "A method for change analysis with weight of significance using multi-temporal, multi-spectral images", in *Proceedings SPIE 2315 European Symposium on Satellite Remote Sensing, Image and Signal Processing for Remote Sensing*, J. Desachy ed., Rome, Italy, pp. 282–288, 1994.

7. H. Hanaizumi and S. Fujimura, "Change detection from remotely sensed multitemporal images using multiple regression", *Proceedings from the 1992 International Geoscience and Remote Sensing Symposium, IGARSS'92*, Houston, Texas, USA, pp. 564–566, 1992.

8. H. Hotelling, "Relations between two sets of variates", *Biometrika*, vol. XXVIII, pp. 321–377, 1936.

9. A. A. Nielsen, *Analysis of regularly and irregularly sampled spatial, multivariate, and multitemporal data*, Department of Mathematical Modelling, Technical University of Denmark, PhD Thesis no. 6, 1994. http://www.imm.dtu.dk/documents/users/aa/phd/.

10. A. A. Nielsen, "Change detection in multispectral, bi-temporal spatial data using orthogonal transformations", *8th Australasian Remote Sensing Conference*, Canberra, Australia, 25–29 March 1996.

11. A. A. Nielsen and K. Conradsen, "Multivariate alteration detection (MAD) in multispectral, bitemporal image data: a new approach to change detection studies", Department of Mathematical Modelling, Technical University of Denmark, Technical Report 1997–11, 1997. http://www.imm.dtu.dk/documents/users/aa/tech-rep-1997-11/.

12. A. A. Nielsen, K. Conradsen and J. J. Simpson, "Multivariate alteration detection (MAD) and MAF post-processing in multispectral, bi-temporal image data: new approaches to change detection studies", *Remote Sensing of Environment*, vol. 64, pp. 1–19, 1998.

13. A. A. Nielsen, R. Larsen and H. Skriver, "Change detection in bi-temporal EMISAR data from Kalø, Denmark, by means of canonical correlations analysis", *Proceedings from the Third International Airborne Remote Sensing Conference and Exhibition*, Copenhagen, Denmark, vol. I, pp. 281–287, 1997.

14. H. Olsson, "Regression functions for multitemporal relative calibration of thematic mapper data over boreal forest", *Remote Sensing of Environment*, vol. 46, no. 1, pp. 89–102, 1993.

15. H. Olsson, *Monitoring of local reflection changes in boreal forests using satellite data*, Ph.D. thesis, Remote Sensing Laboratory, Swedish University of Agricultural Sciences, Umeå, Sweden, 1994.

16. H. Olsson, "Reflectance calibration of thematic mapper data for forest change detection", *International Journal of Remote Sensing*, vol. 16, no. 1, pp. 81–96, 1995.

48   A. A. Nielsen

17. P. Switzer and A. A. Green, "Min/max autocorrelation factors for multivariate spatial imagery", Technical Report no. 6. Department of Statistics, Stanford University, 1984.
18. R. Wiemker, A. Speck, D. Kulbach, H. Spitzer and J. Beinlein, "Unsupervised robust change detection on multispectral imagery using spectral and spatial features", *Proceedings from the Third International Airborne Remote Sensing Conference and Exhibition*, Copenhagen, Denmark, vol. I, pp. 640–647, 1997.

Aspects of Multi-Scale Analysis for Managing Spectral and Temporal Coverages of Space-Borne High-Resolution Images*

Bruno Aiazzi[1], Luciano Alparone[2], Alessandro Barducci[1], Stefano Baronti[1], Roberto Carlà[1], and Ivan Pippi[1]

[1] "Nello Carrara" I.R.O.E. - C.N.R. Via Panciatichi 64, 50127 Firenze, Italy.
e-mail: {baronti,pippi}@iroe.fi.cnr.it
[2] Department of Electronic Engineering, University of Florence, Via S. Marta, 3, 50139 Firenze, Italy.
e-mail: alparone@cosimo.die.unifi.it

Summary. In this work, a multi-resolution procedure based on a *generalized Laplacian pyramid* with *rational* scale factor is proposed to merge image data of any resolution and represent them at any scale. The pyramidal data fusion approach is shown to be superior to a similar scheme based on the discrete *wavelet* transform, according to a set of parameters established in the literature. Not only fused images look sharper than their original versions, but also textured regions are enhanced without losing their spectral signatures. Landsat Thematic Mapper and SPOT Panchromatic images are fused together. The resulting bands capture multispectral features at an increased spatial resolution, thereby expediting automatic analyses for contextual interpretation of the environment.

1. Introduction

The availability of images with higher and higher ground resolutions, collected by satellite sensors having different spectral/temporal coverages, demands the development of novel processing techniques (and improvement of existing ones), to cope with emerging applications requiring advanced modelling, classification and feature extraction. A significant impact might be expected in applications such as small-scale cartography and cadastral map production, as well as in risk or damage assessment (e.g of landslides, floods, earthquakes).

Multi-resolution techniques are extremely attractive for image understanding since they provide a thorough, yet multi-fold, description of the imaged scene. They offer a number of advantages including, but not limited to:

- quick global comprehension of underlying phenomena for simple modelling;
- progressive information parsing through multiple resolutions;
- capability to locally achieve refinements wherever needed;

---

* Work carried out under grants of: CNR -National Research Council of Italy-*Nationwide Project on Cultural Heritage* and of ASI -Italian Space Agency.

50   B. Aiazzi *et al.*

– capability to focus the processing effort on restricted areas at full resolutions.

Wavelets, sub-bands and Gaussian/Laplacian pyramids are among the most suited representations to allow spatial analysis to be carried out on multiple scales. Their primary applications include quality enhancement and de-noising [2], data fusion, textural classification, scene segmentation, and generally all tasks involving iterative relaxation (clustering, matching, etc.).

The goal of this work is to highlight and promote the use of Laplacian Pyramids (LP) as a unifying framework in which a variety of applications requiring high-resolution imagery can be accommodated. In LP, an image is decomposed into nearly-disjoint band-pass channels in the spatial frequency domain, without losing the spatial connectivity of its edges. Versions having lower spatial resolution exhibit larger signal to noise ratios (SNRs), thus allowing robust feature extraction and accurate analysis of image textures. In addition, the LP approach can be easily extended to manage scales whose ratios are rational numbers (not only powers of two, as for frequency-octave wavelet and sub-band decompositions). Thus, data with ground resolutions of, say, 3, 5, and $7m$ can be easily compared and merged (for feature-matching or multi-sensor image fusion). It is possible for LPs to be designed to deal with multi-temporal observations (using spatio-temporal pyramids) to expedite detection of changes occurring at different scales.

2. Multi-Sensor Image Data Fusion

The availability of data from many sensors with different characteristics makes data fusion [7] a topic of ever increasing relevance to the field of remote sensing. The main goals of data fusion are to:

– allow integration of different information sources;
– expedite feature detection and extraction;
– enhance contrast and/or spatial resolution.

The fusion of Landsat Thematic Mapper (TM) and SPOT Panchromatic (PAN) images has already been previously considered in the literature [4] due to their availability and their complementary spatial/spectral characteristics. The recent development of sensors capable of collecting data with extremely high spatial resolutions will open new perspectives for data fusion.

After data resampling to obtain a homogeneous spatial scale, image data fusion may be approached at three different levels:

– **Pixel-level** combination [7]:
  – *deterministic*: Intensity-Hue-Saturation (IHS) transformation;
  – *statistical*: Principal Components Analysis (PCA) based on covariance of the data.
– **Feature-level** combination (e.g. edges, frequency components):

⇒ High-Pass Filtering (HPF) (Chavez *et al.* 1991 [4]).
- **Decision-level** combination of pixel values and/or features.

When processing is performed at a *feature* level, the main objective is to extract significant characteristics and to combine them in such a way as to enhance spatial contrast on the basis of the whole data set [6]. For classification, however, merging algorithms must maintain the spectral characteristics of the original data to avoid misinterpretation. HPF has been found to be far more efficient than the other algorithms in preserving the spectral features of the enhanced bands [4].

The algorithm proposed in the following is a variant of the high-pass filter (HPF) method by Chavez *et al.* [4]. Its generalization is achieved in a pyramid framework, since pyramid structures possess a number of useful and attractive properties. The pyramid-generating filters can be easily designed to cope with images of any resolution from different sensors. Once new data from different sensors become available (e.g. SAR data, digitized aerial photographs, hyperspectral high-resolution aircraft data, data from *new-generation* satellites), they will be easily merged with the existing ones in order to exploit any advantages occurring from a cooperative analysis based on multiple imaging sources.

3. Multiresolution Image Analysis

3.1 Laplacian Pyramid

The Laplacian pyramid is derived from the Gaussian Pyramid (GP) which is a multiresolution image representation obtained through a recursive *reduction* (low-pass filtering and decimation) of the image data set.

Let $G_0(m, n)$, $m = 0, \ldots, M - 1$ and $n = 0, \ldots, N - 1$, $M = u \times 2^K$ and $N = v \times 2^K$, be an image. The classic GP [3] is defined with a decimation factor of 2 ($\downarrow 2$) as

$$G_k(m, n) = reduce_2[G_{k-1}](m, n) \tag{3.1}$$

$$\triangleq \sum_{i=-L_r}^{L_r} \sum_{j=-L_r}^{L_r} r_2(i) \times r_2(j) \, G_{k-1}(2m + i, 2n + j)$$

for $k = 1, \ldots, K$, $m = 0, \ldots, M/2^k - 1$, and $n = 0, \ldots, N/2^k - 1$, in which k identifies the level of the pyramid, K being the top, or *root*, or *base-band*, of size $u \times v$. The 2D reduction *low-pass* filter is given as the outer product of a linear symmetric odd-sized kernel $\{r_2(i)\}$ which should cut-off at one half of the bandwidth of the signal, to prevent *aliasing*.

From the GP, the enhanced LP [1] is defined for $k = 0, \ldots, K - 1$ as:

$$L_k(m, n) \triangleq G_k(m, n) - expand_2[G_{k+1}](m, n) \tag{3.2}$$

in which $expand_2[G_{k+1}]$ denotes the $(k + 1)^{st}$ GP level expanded by 2 to match the size of the underlying k^{th} level:

$$expand_2[G_{k+1}](m,n) \triangleq \sum_{\substack{i=-L_e \\ (j+n) \bmod 2=0 \\ (i+m) \bmod 2=0}}^{L_e} \sum_{j=-L_e}^{L_e} e_2(i) \times e_2(j) \times G_{k+1}\left(\frac{i+m}{2}, \frac{j+n}{2}\right)$$

(3.3)

for $m = 0, \ldots, M/2^k - 1$, $n = 0, \ldots, N/2^k - 1$, and $k = 0, \ldots, K - 1$. The 2D low-pass filter for expansion is given as the outer product of a linear symmetric odd-sized kernel $\{e_2(i)\}$, which again should cut-off at one half of the bandwidth of the signal. Summation terms are taken to be null for non-integer values of $(i + m)/2$ and $(j + n)/2$, corresponding to interleaving zeroes introduced by up-sampling ($\uparrow 2$).

The attribute *enhanced* depends on the fact that the expansion filter is forced to be *half-band*, i.e. an interpolator by 2, and is independent of the reduction filter, which may be *half-band* as well, or not [1].

3.2 Generalized Laplacian Pyramid with Rational Scale Factor

When the ratio between the scales of two images to be merged is not a power of 2, equations (3.2) and (3.3) need be generalized to deal with different factors for reduction and expansion [5]. Reduction by a factor q is defined as

$$reduce_q[G_k](m,n) \triangleq \sum_{i=-L_r}^{L_r} \sum_{j=-L_r}^{L_r} r_q(i) \times r_q(j) \times G_k(qm + i, qn + j) \quad (3.4)$$

The reduction filter $\{r_q(i)\}$ should cut-off at one q^{th} of the signal bandwidth. Expansion by p is defined as

$$expand_p[G_k](m,n) \triangleq \sum_{\substack{i=-L_e \\ (j+n) \bmod p=0 \\ (i+m) \bmod p=0}}^{L_e} \sum_{j=-L_e}^{L_e} e_p(i) \times e_p(j) \times G_k\left(\frac{i+m}{p}, \frac{j+n}{p}\right) \quad (3.5)$$

The *low-pass* filter for expansion $\{e_p(i)\}$ should cut-off at one p^{th} of the signal bandwidth. Summation terms are taken to be null for non-integer values of $(i + m)/p$ and $(j + n)/p$, corresponding to zero samples.

If $p/q > 1$, (where p, q are integer) is the scale factor between two images to be merged, equation (3.2) modifies to the cascade of an *expansion* by q and a *reduction* by p

$$G_{k+1} = reduce_{p/q}[G_k] \triangleq reduce_p\{expand_q[G_k]\} \quad (3.6)$$

while equation (3.3) becomes an expansion by p followed by a reduction by q:

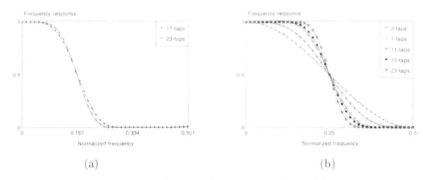

Fig. 3.1. Frequency responses of pyramid-generating filters: (a) *one-third of band*, (b) *half-band*, adopted for reduction and, with an adjusted DC gain, expansion.

$$expand_{p/q}[G_k] \stackrel{\Delta}{=} reduce_q\{expand_p[G_k]\} \quad (3.7)$$

Notice that, when equation (3.4) is cascaded to equation (3.5), convolution for low-pass filtering can be skipped after up-sampling in equation (3.6), as well as before down-sampling in equation (3.7).

The Generalized Laplacian Pyramid (GLP) with p/q scale factor between two adjacent layers, $\tilde{L}_k$, can thus be defined, for $k = 0, \ldots, K-1$, as

$$\tilde{L}_k(m,n) \stackrel{\Delta}{=} G_k(m,n) - expand_{p/q}\{reduce_{p/q}[G_k]\}(m,n) \quad (3.8)$$

while $\tilde{L}_K(m,n) \equiv G_K(m,n)$, and represents a multiresolution image description suitable for merging data imaged with a p/q scale ratio, p and q being two integers prime to each other.

3.3 Filters for Reduction and Expansion

Filters with different frequency responses need to be designed to cope with the scaling requirements of the pyramid algorithms. In particular, for a p/q scale ratio, $p > q$, if normalized frequency is considered, only one filter with $1/p$ cut-off is needed. In the case $p = 3$, two kernels with 17 and 23 taps (filter coefficients) have been designed for pyramid generation. In the case $p = 2$, polynomial kernels with 3 (linear), 7 (cubic), 11 (fifth-order), 15 (seventh-order), and 23 (ninth-order) coefficients have been assessed. The frequency responses of all the filters are plotted in figure 3.1.

The filter design usually stems from a tradeoff between selectivity (sharp cut-off) and computational cost. Instead, the wavelet transform [8] would require also a high-pass filter (i.e., a complete filter-bank) which must generally be re-designed for every value of the ratio p/q.

4. Multiresolution Image Fusion Schemes

4.1 Pyramid-Based Fusion

The block diagram reported in figure 4.1 describes the data fusion algorithm in the general case of two image data sets, initially registered on the same cartographic base, whose scale ratio is p/q.

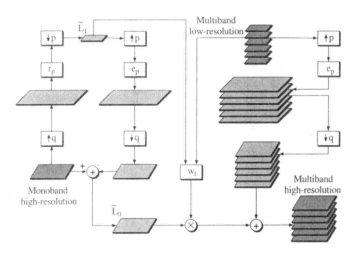

Fig. 4.1. Outline of data fusion procedure for two images, one of which multispectral, with a p/q scale ratio.

Let $S^{(1)}$ be the data set constituted by a single image having smaller scale (i.e. finer resolution) and size $Mp/q \times Np/q$, and $S_l^{(2)}$ $l = 1, \ldots, L$ the data set made up of L multi-spectral observations having scale larger by a factor p/q (i.e., coarser resolution), and thus size $M \times N$. The goal is to obtain a set of L multi-spectral images, each having same the spatial resolution as $S^{(1)}$. The upgrade of $S^{(2)}$ to the resolution of $S^{(1)}$ is the zero-mean GLP (3.8) of $S^{(1)}$, computed for $k = 0$. The images of the set $S^{(2)}$ need only be interpolated by p and then reduced by q to match the finer resolution. After that, the high-pass component from $S^{(1)}$ is added to the expanded versions of $S_l^{(2)}$ $l = 1, \ldots, L$, which constitute the low-pass component, in order to yield a spatially enhanced set of multi-spectral observations, $S_l^{(3)}$ $l = 1, \ldots, L$. Notice that only one level of pyramid decomposition ($K = 1$) is considered when merging two sets of images.

A critical point of the above outline is the equalization of pyramid levels before merging. In fact, the two images may exhibit different *contrast*, but this fact should not affect the results of the fusion. Therefore, the high-frequency component of the image having the higher resolution would better be normalized in such a way that the corresponding *base-band* exhibits the same contrast (intended as average local variance) as any of the lower resolution bands, before being merged.

Let $\tilde{L}_k^{(1)}(m,n)$, $k = 0, 1$ denote the GLP of $S^{(1)}$, and $S_l^{(2)}(m,n)$ the lth band of $S^{(2)}$ both at (m,n), for each band l, a suitable weighting factor w_l is given by

$$w_l = \sqrt{\frac{\sum_{m=0}^{M-1}\sum_{n=0}^{N-1} var[S_l^{(2)}(m,n)]}{\sum_{m=0}^{M-1}\sum_{n=0}^{N-1} var[\tilde{L}_1^{(1)}(m,n)]}} \qquad (4.1)$$

in which $var[\cdot]$ denotes local variance on a 7 × 7 neighbourhood. Thus, the high-frequency component of $S^{(1)}$ will be multiplied by w_l before being added to the expanded version of $S_l^{(2)}$.

4.2 Wavelet-Based Fusion

The idea of the wavelet-based fusion algorithm developed by Li et al. [6] is to merge couples of sub-bands of corresponding frequency content on the basis of an activity measure locally computed on 2 × 2 blocks of coefficients. The fused image is produced by taking the inverse transform of the blocks of coefficients chosen as the more *active* between the two images. Figure 4.2 outlines the fusion procedure extended to three images having different scales. The middle frequency coefficients are selected from either of the low or medium scale, based on their local variance. All coefficients except those on the *base-band* are weighted, as shown above.

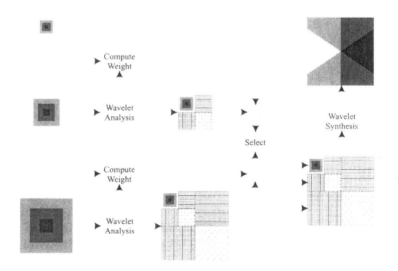

Fig. 4.2. Outline of wavelet-based image fusion procedure for three images, whose scale ratios are powers of two.

5. Experimental Results

5.1 A Comparison of Wavelet and Pyramid Image Fusion

The Landsat TM Band 6 image of figure 5.1(a) was merged with the remaining six bands. Figure 5.1 shows the results relative to Band 5 (figure 5.1(b)). Due to their extremely low energy, the six higher frequency sub-bands of Band 6 have been straightforwardly replaced with those of Band 5. Although the wavelet-fused image looks sharper, artifacts are perceivable around edges, due to ringing effects. Also "striping" effects, periodic patterns of horizontal stripes due to imperfect calibration of the 16 imaging sensors of TM, are less annoying in the pyramid-fused image.

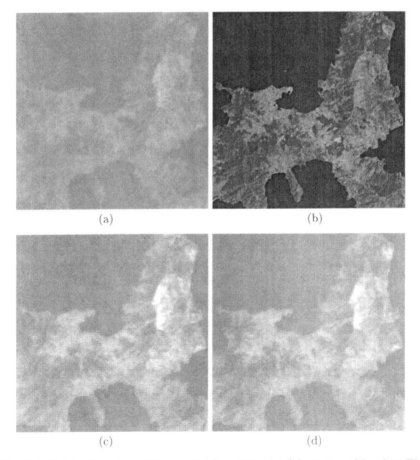

Fig. 5.1. Original Landsat TM Band 6 (a) and Band 5 (b). Fusion of Landsat TM Band 6 with Band 5: (c) wavelet scheme; (d) pyramid scheme.

Spectral feature preservation is evaluated by taking the pixel differences between any of the merged images and a linearly resampled version of Band 6 (both integer valued). These differences are expected to be either zero or very small on homogeneous areas, and relevant on contours or highly textured areas. The standard deviation of such differences and the number of pixels in which they are equal or very close to zero represent two figures of merit for image data fusion [4]. Standard deviations should be as close to zero as possible. Due to rounding to integers, pixel differences are taken to be null if their absolute values do not exceed one. Null differences should indicate the fraction of pixels belonging to homogeneous areas.

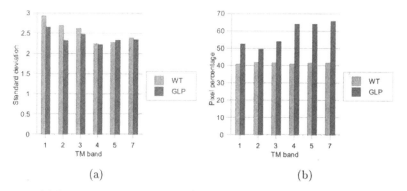

Fig. 5.2. (a) Standard deviations (STD) of the differences generated by subtracting the merged images from the corresponding TM bands and (b) Percentage of pixels ($P \pm 1$) whose absolute differences between the merged TM and SPOT images and the corresponding expanded TM band are equal to either one or zero. Results reported for both wavelet (WT) and pyramid (GLP) schemes.

Figure 5.2 shows standard deviation values and the percentage of pixels being close by no more than one, for each TM band and for the two different fusion schemes. It is clear that the pixel percentages are far larger for the pyramid scheme. Standard deviations are slightly smaller for the pyramid, with the only exception of Band 5. It is apparent that performances are better for the pyramid algorithm than for the wavelet-based one. The scores are similar for each "infrared" TM Band (4, 5, and 7) and inferior for TM Bands in the visible wavelengths (1, 2, and 3), due to better homogeneity of the data to be merged. The values of the parameters reported have been optimized over the pyramid-generating filter (15 taps) and are extremely steady, unless poor filters are employed, and so close to their asymptotic values that no significant improvement can be expected by designing more sophisticated filters.

The pyramid scheme outperforms the wavelet scheme by a significant extent, notwithstanding contrast equalization, which was missing in the original paper [6] but was applied also to the wavelet scheme to improve its scores.

58   B. Aiazzi *et al.*

Such a behaviour might be attributed to the characteristics of wavelet sub-bands. In fact, a higher-frequency sub-band contains the aliasing contribution of the lower-frequency sub-band, and vice-versa. Therefore, sub-band replacement from one image to another will introduce *noise-like* aliasing artifacts after image synthesis. Instead, GLP does not require a perfect reconstruction filter bank to ensure aliasing compensation.

5.2 A Case Study

The study area of Metaponto, of which SPOT Panchromatic and Landsat TM data were available, is located in Southern Italy (Lucania region) and spans about 1300 km^2. The images were registered on the same cartographic base, each maintaining its own scale. A first-order polynomial model was used together with cubic resampling. The accuracy of registration obviously affects the performances of the fusion procedure. The proposed pyramid algorithm was applied to the test area. For the case addressed, p/q ratio is 3. This simplifies the outline as no reduction/expansion by q is necessary ($p = 3$, $q = 1$). Visual results (inc. original SPOT-PAN and Landsat TM Band-5) are shown only for a small portion of the test area, of about 2.5×2.5 km^2 (figure 5.3). Contours and textures are highlighted. In fact, since image texture is related to spatial frequency components, the band-pass approach of GLP is likely to preserve its main features, without introducing the typical artifacts encountered in wavelet-based fused images. The local average level, however, is carefully preserved. Such a feature is important in determining spectral signatures, and its alteration may be responsible for misclassification and misinterpretation.

Figure 5.4 reports Chavez's parameters for fusion of all TM bands with SPOT. Even though the magnitude of such scores is image-dependent, they can be easily assessed for each single image. Compared with the original HPF method employing 5×5 box filters, the pyramid framework allows a far better performance to be achieved with filters relatively simple and easy to design. The additional benefit produced by 23-taps filters over 17-taps filters is negligible compared to the large advantage of pyramid filters over box filters. Results are superior mainly thanks to the multi-resolution pyramid framework which allows a better filter design.

6. Conclusions

The novel procedure described here generalizes the *feature-based* image fusion of HPF in a pyramid context. It allows a looser choice of image scales, no longer constrained to be multiples of one another, and retains multispectral signatures far better, according to scores in experimental tests.

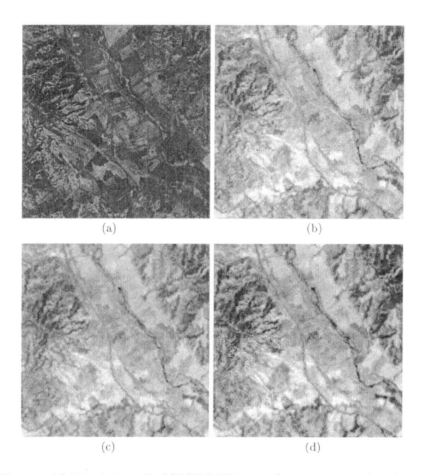

Fig. 5.3. (a) 510 × 510 detail of SPOT-PAN image of test area, ground resolution 10m, (b) 170 × 170 TM-5 image of test area, resolution is 30m and a magnification by 3 has been applied for displaying. TM-5 image fused with SPOT-PAN at 30m scale: (c) pyramid method and (d) HPF.

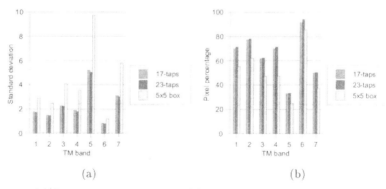

Fig. 5.4. (a) Standard deviations and (b) percentage of pixels, defined as in figure 5.2, for box filter and for the two filters shown in figure 3.1(a).

References

1. B. Aiazzi, L. Alparone, S. Baronti, and F. Lotti, "Lossless image compression by quantization feedback in a content-driven enhanced Laplacian pyramid," *IEEE Transactions on Image Processing*, vol. 6, no. 6, pp. 831–843, 1997.
2. B. Aiazzi, L. Alparone and S. Baronti, "Multiresolution local-statistics speckle filtering based on a ratio Laplacian pyramid," *IEEE Transactions Geoscience and Remote Sensing*, vol. 36, no. 5, pp. 1466–1476, 1998.
3. P. J. Burt, "The pyramid as a structure for efficient computation," in *Multiresolution Image Processing and Analysis*, A. Rosenfeld (ed.), Springer, 1984.
4. P. S. Chavez Jr., S. C. Sides, and J. A. Anderson, "Comparison of three different methods to merge multiresolution and multispectral data: Landsat TM and SPOT panchromatic," *Photogrammetric Engineering and Remote Sensing*, vol. 57, no. 3, pp. 295–303, 1991.
5. M. G. Kim, I. Dinstein, and L. Shaw, "A prototype filter design approach to pyramid generation," *IEEE Transactions on Pattern Analysis and Machine Intelligence*, vol. 15, no. 12, pp. 1233–1240, 1993.
6. H. Li, B. S. Manjunath, and S. K. Mitra, "Multisensor image fusion using the wavelet transform," *Computer Vision Graphics and Image Processing: Graphical Models and Image Processing*, vol. 57, no. 3, pp. 235–245, 1995.
7. R. C. Luo and M. G. Kay, "Multisensor integration and fusion in intelligent systems," *IEEE Transactions on Systems, Man, and Cybernetics*, vol. 19, no. 5, pp. 901–931, 1989.
8. S. Mallat, "A theory for multiresolution signal decomposition: the wavelet representation," *IEEE Transactions on Pattern Analysis and Machine Intelligence*, vol. 11, no. 7, pp. 674–693, 1989.

Structural Inference Using Deformable Models

Jens Michael Carstensen, Rune Fisker, Nette Schultz, and Torsten Dörge

Department of Mathematical Modelling, Technical University of Denmark, Building 321, DK-2800 Denmark.
e-mail: jmc@imm.dtu.dk

Summary. Deformable models have proved useful in many machine vision applications during the last decade. To increase performance and to make their use computationally feasible they have to be specifically tuned for the application domain. Model identification and parameter estimation are receiving increasing attention. The aim of the paper is by no means to give a thorough treatment of the theory behind deformable models, but to illustrate their merits in three practical applications using different types of deformable models: segmenting nano particles, hybridization filter analysis, and shape analysis in the meat industry. Although not directly developed with remote sensing applications in mind analogous models may be useful in this area.

1. Bayesian Image Analysis

Bayesian image analysis is a framework for incorporating stochastic models of visual phenomena into image analysis. The Bayesian paradigm in image analysis can be described as follows [1]:

1. Construct a prior probability distribution $P(v)$ for the visual phenomena V, that we want to make inference about.
2. Formulate an observation model $P(y|v)$. This is the distribution of observed images Y given any particular realization v of the prior distribution.
3. Combine the prior distribution and observation model to the posterior distribution $P(v|y)$ by Bayes theorem.

$$P(v|y) \propto P(v)P(y|v) \tag{1.1}$$

   $P(v|y)$ is the distribution of the visual phenomena V given the images y that we have observed.
4. Make inference about the visual phenomena based on the posterior distribution $P(v|y)$.

For the Bayesian approach to be successful it is important that the prior density reflects our knowledge about the visual phenomena.

2. Deformable Models

Deformable models is a powerful tool in the framework of Bayesian image analysis when the task is to make inference about shape and structure. The

62   J. M. Carstensen *et al.*

basic idea of deformable models is that a structure embedded in the image can be considered as a deformation of a given template uniquely represented by the template parameters v. The prior distribution represents the prior knowledge about the structure, i.e. the deformable template.

Deformable templates can roughly be separated into 2 groups [9]: Free form and Parametric. Free form deformable templates have no explicit global structure because the prior only contains local continuity and smoothness constraints. This makes free form templates able to represent an arbitrary shape as long as the continuity and smoothness constraints are satisfied. The best known example of free form deformable templates are active contour models (snakes) proposed by [10] and further developed by e.g. [3, 12].

In parametric deformable models prior knowledge of the global structure is included using a parameterized template. Grenander *et al.* [8] present a deformable model, which consists of a prototype template and a probabilistic distribution of the deformation space. Jain *et al.* [9] present a general framework for object matching using deformable templates, which is also based on a prototype of the given structure and local constraints on the variability. Garrido *et al.* [6] present a model very similar to the model described in [9], but use different algorithms for initialization and optimization. The deformable models presented by [6, 8, 9] can all handle an arbitrary structure, as long as the prototype template is specified. Most other parametric deformable models are formulated for a specific structure. Yuille *et al.* [14], for example, use an eye and a mouth model to locate the eyes and mouth in face images, respectively, Carstensen [2] models hybridization filters to investigates images from genetic experiments and Fisker and Carstensen [4] use yarn models to perform visual inspection of textile.

In most of the deformable models presented in the literature the prior model $P(v)$ consists of a quadratic term, which penalize the deviation of a given model configuration with respect to the ideal configuration. The quadratic term gives the opportunity to simulate the prior model by sampling in the conditional or unconditional prior distribution [5]. In many cases the observation model is formulated so that the interaction between a realization of the template v and an observed image y corresponds to image intensity [2, 4, 8, 14] and/or edge information [3, 6, 9, 10, 12, 14], but in principle all kind of information can be combined e.g. texture or colour. Another characteristic of deformable models is that a number of model parameters, which give the relative influence of different terms in the model, have to be selected. Most authors select the model parameters based on empirical observations. A method for supervised and unsupervised estimation of the parameters is presented in [5].

The final step in the Bayesian paradigm for image analysis is to make inference based on the posterior distribution $P(v|y)$. This is done e.g. by estimating the maximum a posteriori (MAP) defined as:

$$\hat{v} = \arg\max_{v} P(v|y) \tag{2.1}$$

The MAP-estimation is a high-dimensional optimization problem. Many different techniques have been applied for MAP estimation: deterministic [3, 6, 9, 10, 12, 14], stochastic [2, 8] and heuristic optimization algorithms [4]. Commonly used optimization algorithms demand an initial configuration of the template parameters v to be specified. Most authors define the initial configuration manually or using an ad hoc method. General methods for initialization have also been presented, which are based on a modified Hough transform [6], a multi-resolution approach [9] and on the generalized Hough transform [12].

3. Examples

This section presents three real world examples of the formulation and use of deformable models.

3.1 Nano Particles

In the upper left corner of figure 3.1 we show a close-up of a scanning electron microscope image of nano particles. These nano particles are known to be approximately elliptical. The presence of significant noise makes direct filtering approaches end up with highly irregular objects representing the particles. We want to impose the elliptical shape on the particles by using a deformable ellipsis model. The parameters of the detected ellipses will then capture the relevant size and shape characteristics of the particles.

In this study we assume knowledge on the expected size of the particles or that this can be set as a turn-button parameter. Automated procedures for this should be fairly easy to construct. A circular filter with the same size is convolved with the image and the result is shown in figure 3.1, upper right. The filtered image is then thresholded and labelled, creating a marker for each particle in the image (see figure 3.1, lower left). This image can now be used to initialize the positions of the deformable ellipses.

A particle is modelled by an ellipse with the template parameters

$$v = (r_0, c_0, a, b, \theta)$$

where (r_0, c_0) is the ellipse centre, a and b are the ellipse axes and θ is the rotation. As a crude assumption the template parameters are set to be uniformly distributed i.e. a uniform – and improper – prior model. The observation model is given by:

$$P(y|v) = \tfrac{1}{Z_o} \times \exp\{-(\tfrac{1}{|\Omega_{in}|} \textstyle\sum_{(r,c)\in\Omega_{in}} I(r,c) - \tfrac{1}{|\Omega_{out}|} \sum_{(r,c)\in\Omega_{out}} I(r,c))\} \tag{3.1}$$

where,

$$\Omega_{in} = \{(r,c) | \frac{(r_r)^2}{a^2} + \frac{(c_r)^2}{b^2} \leq 1\} \text{ and,}$$

$$\Omega_{out} = \{(r,c) | 1 < \frac{(r_r)^2}{a^2} + \frac{(c_r)^2}{b^2} \leq 2\}$$

where,

$$r_r = (r - r_0)cos(\theta) + (c - c_0)sin(\theta) \text{ and,}$$
$$c_r = (c - c_0)cos(\theta) - (r - r_0)sin(\theta)$$

$I(r,c)$ is the image intensity at row r and column c and Z_o is the normalizing constant. The interpretation of the observation model is that the pixels in a band around the ellipse are subtracted from the pixels within the ellipse, where the band have the same area as the area of the ellipse. The probability is then increased by maximizing the difference.

As an initial configuration of (r_0, c_0) and θ for a particle, is used the centre of gravity and the major orientation of the corresponding marker. Initially the particle is assumed to be circular, i.e. $a = b$, with the previously given size. Finally simulated annealing is used to do the MAP-estimation. The particles in figure 3.1, lower right, are located using this scheme.

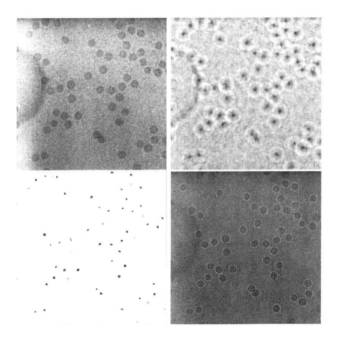

Fig. 3.1. Upper left: Original close-up image of nano particles. Upper right: Original after a matched filtering. Lower left: Labelled markers for nano particles. Lower right: Estimated contours overlayed on original.

3.2 Hybridization Filters

Hybridization filters are used for high-throughput screening of cosmid clones – stretches of DNA app. 40.000 bases long – in e.g. the human genome project. The image analysis problem is to locate spots placed in a rectangular lattice by a robot. Several circumstances complicate the localization problem: imprecise robot movements, geometric warp, missing spots, merged spots etc. The localization problem has been addressed in [2].

A good quality hybridization filter is shown in figure 3.2. The lattice structure is parameterized using the template parameters $v = \{(r_i, c_i), i = 1, \ldots, N_n\}$ that contains the (r, c) image position of the nodes in the lattice. The prior distribution imposed on v is defined as:

$$P(\mathbf{v}) \propto \exp\{-\alpha_0 \sum_{i \sim j} (\frac{d_0(i,j)}{D})^2 - \alpha_1 \sum_{i \sim j} (\frac{d_1(i,j)}{D} - 1.0)^2\} \quad (3.2)$$

where i and j represent nodes and $i \sim j$ means that nodes i and j are nearest-neighbours. $d_0(i,j)$ is the deviation in alignment of the nodes i and j, and $d_1(i,j) - D$ is the deviation from the fixed lattice distance, D, between neighbouring nodes. Figure 3.3 illustrates the meaning of d_0 and d_1. For the nodes on the edges we use the distance to the outline instead of the distance to the non-existing neighbour. α_1 and α_2 are model parameters.

Given the spot locations **v** we then specify an observation model for the observed image **y** as

$$P(\mathbf{y}|\mathbf{v}) \propto \exp(\frac{1}{\lambda} \sum_i \mu(i))$$

Fig. 3.2. Good quality 96 × 96 hybridization filter.

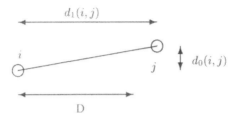

Fig. 3.3. Definition of the distance measures d_0 and d_1 for a horizontal neighbour-pair. D is the distance between neighbours on a perfect lattice. The definition for vertical neighbour-pairs is analogous.

where the summation is over all spots i, $\mu(i)$ is the sum of the gray levels in a 5×5 neighbourhood around spot i and λ is the regularization parameter, which gives the relative influence of the observation model.

Figure 3.4 shows the results of the spot localization using the deformable model. Even for very noisy images the spot lattice is correctly found.

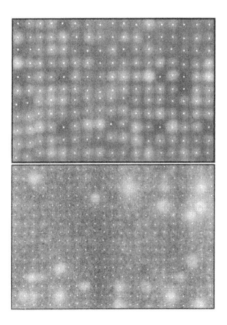

Fig. 3.4. Two localization examples. On the top a good quality image where two rows merge and split up again. On the bottom a noisy image where most spots are hardly noticeable.

3.3 Inspection of Meat

This example concerns the estimation of meat percent in pork carcasses. Given two cross-sectional images – one at the front and one near the ham of the carcass – the areas of lean and fat should be measured automatically.

This work is based on the deformable model proposed in [8]. We refer to [8, 11, 13] for a thorough exposition. This presentation should rather be seen as an introduction. The proposed template consists of a 2D vector cycle $v = (v_0, ..., v_{n-1})$ where v satisfy the closure conditions $\sum_{i=0}^{n-1} v_i = 0$. To locate the relevant structure a global transformation T followed by a local transformation S is applied to the initial template v^0:

$$v^0 \xrightarrow{T} v^T \xrightarrow{S} v^S \tag{3.3}$$

where the initial template v^0 is a general shape model of the actual structure. The transformations is performed by changing the orientation and length of a set of the vectors. This is done by updating each vector $v_i^t = S_i(\gamma_i, \phi_i) v_i^{t-1}$ where $S_i(\gamma_i, \phi_i)$ is the transformation matrix for v_i. Matrices changing the length and orientation are:

$$\begin{bmatrix} \rho & 0 \\ 0 & \rho \end{bmatrix} , \quad \rho > 0 \quad \text{and} \quad \begin{bmatrix} \cos \psi & \sin \psi \\ -\sin \psi & \cos \psi \end{bmatrix} , \quad \psi \in [-\pi, \pi]. \tag{3.4}$$

where ρ and ψ specify the change in length and rotation, respectively. Their product can be written in the following linear form:

$$S(\gamma, \phi) = \begin{bmatrix} 1 + \gamma & -\phi \\ \phi & 1 + \gamma \end{bmatrix} \tag{3.5}$$

where,

$$\gamma = \rho \cos(\psi) - 1 \quad \text{and} \quad \phi = -\rho \sin(\psi) \tag{3.6}$$

γ and ϕ specify changes from the identity. For small values of ψ and ρ near 1, $\gamma \approx \rho - 1$ controls changes in length, and $\phi \approx -\psi$ controls changes in orientation. To keep the template vector cycle closed a closure constraint has to be introduced.

Grenander *et al.* [8] assume that the parameter cycles $\gamma = (\gamma_0, \ldots, \gamma_{n-1})^T$ and $\phi = (\phi_0, \ldots, \phi_{n-1})^T$ follow independent central cyclic Markovian Gaussian densities. The cyclic Markovian density for γ can be factorized as:

$$f(\gamma) = \frac{1}{Z_\gamma} \exp\{-\frac{1}{2} \sum_{i=0}^{n-1} (\alpha_{1,i}^\gamma (\gamma_{i+1} - \gamma_i)^2 + \alpha_{2,i}^\gamma (\gamma_i)^2)\} \tag{3.7}$$

where Z_γ is a normalization constant, $\alpha_{1,i}^\gamma \in R_+$ and $\alpha_{2,i}^\gamma \in R_+$ are prior parameters, which give the relative weight of each term. The first term in the exponent of (3.7) controls the bond of γ_i between neighbours and the second the variability of γ_i. An equivalent density $f(\phi)$ can be formulated for ϕ.

68   J. M. Carstensen *et al.*

The global transformation T changes the full template in scale, rotation and horizontal and vertical direction. In the global transformation [8] assume the change in length γ and the changes in rotation ϕ to be the same for all vectors v_i, i.e. $\phi_i = \phi$ and $\gamma_i = \gamma$. Using (3.7) and this assumption the prior density for the global transformation is defined as:

$$P_T(v^{t+1}|v^0) = f(\gamma)f(\phi) = \frac{1}{Z_\gamma Z_\phi} \exp\{-\frac{1}{2}\sum_{i=0}^{n-1}(\alpha_{2,i}^\gamma(\gamma_i)^2 + \alpha_{2,i}^\phi(\phi_i)^2)\} \quad (3.8)$$

where γ_i and ϕ_i are obtained solving $v_i^t = S_i(\gamma_i, \phi_i)v_i^0$. See [8] for the final prior model $P_T(v^{t+1}|v^0)$ where the γ_i and ϕ_i substitution is eliminated. Note the difference between the assumption $\phi_i = \phi$ and $\gamma_i = \gamma$ and the generation of v_i^{t+1} from v_i^0, which leads to the actual values of γ_i and ϕ_i. The generation of v_i^{t+1} from v_i^0 depends on the chosen update strategy, i.e. optimization algorithm. What makes the scheme proposed by [8] unique, but also complex, is that the update v^{t+1} is generated by sampling in the conditional posterior distribution $P_T(v^{t+1}|y, v^0)$. It is beyond the scope of this article to derive the final conditional distributions. For further information we refer to [8, 11, 13].

In the local transformation S only a subset of m vectors is updated per iteration. Grenander *et al.* [8] assume that each vector v_i is updated by individual γ_i and ϕ_i. Using (3.7) and this assumption, the prior density for the local transformation is defined as:

$$P_S(v^{t+1}|v^T) = \frac{1}{Z_\gamma Z_\phi} \exp\{-\frac{1}{2}\sum_{i=0}^{n-1} (\alpha_{1,i}^\gamma(\gamma_{i+1} - \gamma_i)^2 + \alpha_{2,i}^\gamma(\gamma_i)^2 +$$

$$\alpha_{1,i}^\phi(\phi_{i+1} - \phi_i)^2 + \alpha_{2,i}^\phi(\phi_i)^2)\} \quad (3.9)$$

where γ_i and ϕ_i are obtained solving $v_i^t = S_i(\gamma_i, \phi_i)v_i^T$. [8] also updates v^{t+1} in the local transformation by sampling in the conditional posterior distribution $P_S(v^{t+1}|y, v^T)$, which is expanded in [8] to handle the closure constraints and eliminate γ and ϕ. In [8] the observation model is defined as:

$$p(y|v) = \frac{1}{Z_o} \exp(-\frac{1}{2\tau^2}[\ \sum_{(r,c)\in S_{in}}(I(r,c) - m_{in})^2 +$$

$$\sum_{(r,c)\in S_{out}}(I(r,c) - m_{out})^2] \quad (3.10)$$

where S_{in} is the set of pixels belonging to the template, S_{out} is the set of pixels in the image that don't belong to the template, m_{in} is the mean of S_{in}, m_{out} is the mean of S_{out}, τ^2 is the weight variance

$$\tau^2 = \frac{1}{N_{pixel}}[\sum_{(r,c)\in S_{in}}(Ir,c - m_{in})^2 + \sum_{(r,c)\in S_{out}}(Ir,c - m_{out})^2],$$

Z_o is the normalization constant and $I(r,c)$ is the image intensity at row r and column c.

An example of the use of the template, to segment the meat, is shown in figure 3.5. As optimization method is used a simulated annealing scheme equivalent to [7]. For further information on the segmentation see [13].

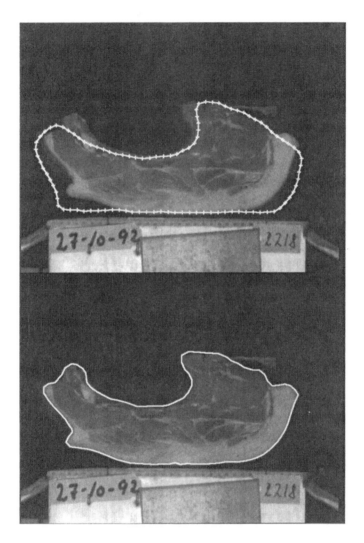

Fig. 3.5. Segmentation of cut on a front cross-section image. The top image shows the start template and the bottom image shows the final deformed template.

70 J. M. Carstensen *et al.*

4. Discussion

Deformable models are a very powerful image analysis tools, and they have already proved useful in a number of applications in the field of machine vision. We have shown the technique applied in three real world problems. Although these problems are not within the field of remote sensing we believe that there are many situations where they can prove useful in this community especially when resolution of remotely sensed images increases and shape and structure gain importance as visual cues. The models similar to those used in the nano particle and meat inspection examples can e.g. be used when detecting roads and buildings. The model in the hybridization filter analysis could prove useful e.g. when counting trees in plantations or whenever a lattice pattern occurs on the ground.

Acknowledgement. The nano particle images were supplied by the Mösbauer group at the Department of Physics, Technical University of Denmark. The hybridization filter images were supplied by Imperial Cancer Research Fund, London, UK. The meat images were supplied by the Danish Meat Research Institute, Roskilde, Denmark.

References

1. J. Besag, "Towards bayesian image analysis", *Journal of Applied Statistics*, vol. 16, no. 3, pp. 395–407, 1989.
2. J. M. Carstensen, "An active lattice model in a bayesian framework", *Computer Vision and Image Understanding*, vol. 63, no. 2, pp. 380–387, 1996.
3. L. D. Cohen, "On active contour models and balloons", *Computer Vision Graphics and Image Processing (CGVIP): Image Understanding*, vol. 53, no. 2, pp. 211–218, 1991.
4. R. Fisker and J. M. Carstensen, "Automated visual inspection of textile", in: *Proceedings of The 10th Scandinavian Conference on Image Analysis*, (Lappeenranta, Finland), pp. 173–179, 1997.
5. R. Fisker and J. M. Carstensen, "On parameter estimation in deformable models", in *Proceedings of the 14th International Conference on Pattern Recognition*, (Brisbane, Australia), vol. 1, pp. 763–766, 1998.
6. A. Garrido and N. Perez De La Blanca, "Physically-based active shape models: Initialization and optimization", *Pattern Recognition*, vol. 31, pp. 1003–1027, 1998.
7. S. Geman and D. Geman, "Stochastic relaxation, gibbs distributions and the bayesian restoration of images", *IEEE Transaction on Pattern Analysis and Machine Intelligence*, vol. 6, pp. 721–741, 1984.
8. U. Grenander, Y. Chow, and D. M. Keenan, *Hands: A pattern theoretic study of biological shapes*, Springer-Verlag, 1991.
9. A. K. Jain, Y. Zhong, and S. Lakshmanan, "Object matching using deformable templates", *IEEE Transactions on Pattern Analysis and Machine Intelligence*, vol. 18, no. 3, pp. 267–278, 1996.

10. M. Kass, A. Witkin, and D. Terzopoulos, "Snakes: Active contour models", *International Journal of Computer Vision*, vol. 8, no. 2, pp. 321–331, 1988.
11. A. Knoerr, *Global models of natural boundaries: Theory and applications.* PhD thesis, Brown university, Providence, Rhode Island, 1988.
12. K. F. Lai and R. T. Chin, "Deformable contours: Modeling and extraction", *IEEE Transactions on Pattern Analysis and Machine Intelligence*, vol. 17, no. 11, pp. 1084–1090, 1995.
13. N. Schultz and K. Conradsen, "2d vector cycle deformable templates", *IEEE Signal Processing*, To appear.
14. A. L. Yuille, P. W. Hallinan, and D. S. Cohen, "Feature extraction from faces using deformable templates", *International Journal of Computer Vision*, vol. 8, no. 2, pp. 99–111, 1992.

Terrain Feature Recognition Through Structural Pattern Recognition, Knowledge-Based Systems, and Geomorphometric Techniques

Dimitris P. Argialas

Remote Sensing Laboratory, Department of Rural and Surveying Engineering,
National Technical University of Athens,
9 Heroon Polytechniou St., Zographos 15780, Greece.
e-mail: argialas@central.ntua.gr

Summary. This chapter reports on a set of four developments, which have resulted in prototype computer software systems related to terrain analysis and interpretation based on techniques from advanced image processing and machine vision. The first is the computational description and identification of drainage patterns through structural pattern recognition. The second uses a variety of expert system methods and tools to address terrain knowledge representation and to construct prototype expert systems for inferring both landform and the physiographic region of a site from user observations of indicators. The third concerns a "terrain visual vocabulary" based on a Macintosh hypermedia system consisting of interlinked definitions, graphics, and aerial images which can be used simultaneously with expert systems to assist novice interpreters. The fourth development concerns a geomorphometric approach for the classification of the GTOPO30 digital elevation model into three classes of physiographic features and for the identification, representation, and classification of mountain objects.

1. Introduction

Terrain analysis can be performed through the systematic study of image elements relating to the nature, origin, morphologic history and composition of distinct units called landforms [10, 13, 19]. Landforms are natural terrain units which, when developed under similar conditions of climate, weathering, and erosion, exhibit a distinct and predictable range of visual and physical characteristics. The entity of a landform is fundamental in representing and organizing topographic and geomorphic information through the pattern element approach to terrain analysis. The pattern elements examined include topographic form, drainage pattern, gully characteristics, soil tone variation and texture, land use, vegetation, plus other special features. The landform is inferred from the pattern elements of the site and then the parent material is inferred by its association with the landform. This approach has wide applications in civil engineering, soil science, environmental inventory and mapping, in agricultural engineering, in structural geology, and in hazard and risk modelling [14, 17, 19]. Many aspects of structural pattern analysis in environmental images described here are analogous to those required in machine vision and related disciplines.

Terrain analysis can be time consuming, labour intensive and costly. It normally requires skills which are acquired through lengthy and expensive training. Therefore, it could be helpful to at least partially automate this process by developing computer-assisted interactive systems [2]. It should be noted that landforms and pattern elements, including drainage patterns, are relatively poorly described components of landscapes and very few of these features can be extracted automatically by standard image analysis or geomorphometric techniques.

This paper reports on various research efforts, which have each produced an original, practical prototype computer system applied to terrain interpretation. Each one is described below.

2. Structural Drainage Pattern Recognition

A drainage pattern is the configuration or shape of a set of tributaries within a drainage network. Comprehensive empirical descriptions of more than thirty pattern configurations were discussed by Howard [9]. Drainage patterns are associated with topographic form, land use, soil type, rock type, lithologic type and geologic structure. Drainage patterns are used in geology, geomorphology, and remote sensing because they are useful for the recognition of the landforms and structures of a region. Computer classification of drainage patterns is useful for formalizing and automating the classification of textures and structures appearing in remotely-sensed images.

Argialas [2] and Argialas and Roussos [3] developed a prototype software system for the identification of drainage patterns by a structural pattern-recognition approach. The Drainage Pattern Analysis (DPA) system was originally programmed on a mainframe computer and later as DPA-PC, within the Microsoft Windows environment. Structural pattern recognition for drainage patterns included: the design of conceptual drainage pattern models and their corresponding numerical hierarchical and relational models, selection and extraction of pattern attributes, and design of a classification strategy. Patterns were described and classified by modelling topological, geometric, and structural relationships among constituent streams. The drainage pattern hierarchy was composed of semantic objects, Strahler segments, reaches, and nodes. Nodes were aggregated to reaches, reaches to Strahler segments, and Strahler segments to semantic objects. An attribute list was designed and attached to each node of the object hierarchy, in the form of a relational table, to characterize the object of that node. Figure 2.1 shows examples of drainage pattern recognition through the DPA-PC system. Figure 2.2 shows computed angle statistics of a drainage pattern. It is concluded that structural pattern recognition provides a formal framework for describing the structure of drainage patterns.

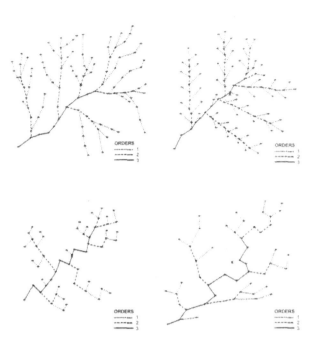

Fig. 2.1. Drainage pattern recognition through the DPA-PC system. Clockwise from upper left dendritic, pinnate, rectangular, and angular patterns [2].

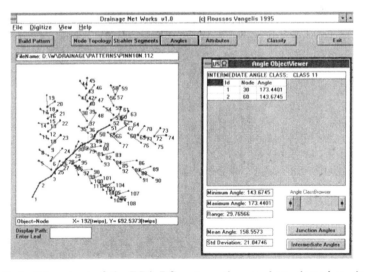

Fig. 2.2. In this output of the DPA-PC system, the user has selected to view the computed angle statistics (min, max, mean, standard deviation, range) in particular those related to the intermediate angles of this pinnate drainage pattern. This pattern has been recognized as pinnate by the system [3].

3. Knowledge-Based Landform Interpretation

Recent advances have indicated that expert systems can be used to capture the knowledge and underlying expertise relevant to many tasks in remote sensing [1]. The expert system approach to terrain analysis problem solving was implemented with production rules involving inexact reasoning, frames, and fuzzy sets [2, 13]. The systems described were termed Terrain Analysis eXperts (TAX-1, 2, 3, 4, 5).

In TAX-1 factual knowledge described landforms in relation to their pattern elements and physiographic sections in relation to their expected landforms. Strategic knowledge (problem-solving decisions) were represented by inexact production rules through a Bayesian formalism [2]. Based on user responses to queries about the physiographic section of the site, the system constructed a set of candidate landforms of the site and estimated their a priori probabilities. TAX then chose the landforms in this candidate list, one by one, and attempted to establish each one of them, by matching the user-supplied pattern-elements of the site with those expected for each landform. A typical consultation script generated with the terrain analysis expert system TAX-1 is shown below. The numeric responses of the user, between -3 and 3, indicate the user's certainty for the presence of the specific pattern-element value in the study area.

Please provide the following information about the site:
To which Physiographic-section does the site belong? Cumberland-plateau
    Is the "gully-amount" of the site "none" ? -3
    Is the "gully-amount" of the site "few" ? 1
    Is the "gully-type" of the site "v-shaped" ? 3
    Is the "landuse-valleys" of the site "cultivated" ? -1
    Is the "landuse-valleys" of the site "forested" ? 3
    Is the "landuse-slopes" of the site "cultivated" ? -3
    Is the "landuse-slopes" of the site "forested" ? 3
    Is the "soil-tone" of the site "medium" ? 1
    Is the "soil-tone" of the site "light" ? 0
    Is the "soil-tone" of the site "dark" ? 0
    Is the "drainage-texture" of the site "coarse" ? 3
    Is the "drainage-type" of the site "internal" ? -2
    Is the "drainage-type" of the site "angular" ? 2
    Is the "topography" of the site "steep-slopes" ? 3
    Is the "gully-amount" of the site "many" ? -2
The site appears to be "sandstone-humid"
The certainty associated with this result is "0.99".

The Terrain Analysis Expert-2 (TAX-2) system was designed in the Intelligence Compiler, a frame and rule based expert-system tool. TAX-2 demonstrated the representation and reasoning capabilities of frames, backward and forward chaining rules, and inexact reasoning for landform interpretation. The Terrain Analysis Expert-3 (TAX-3) system was designed to represent

76   D. Argialas

the vagueness and imprecision that is inherent in the qualitative descriptions of terrain terms by fuzzy sets. Fuzzy set approaches provide a way of dealing with vague linguistic descriptions such as "gentle relief", and "partly dendritic, partly rectangular drainage pattern". For example, a linguistic label such as "Gentle Relief" may be construed as a fuzzy restriction on the values of the base variable "Relief in feet". Thus a flat plain or a site having a Relief of 0m would be definitely called Gentle Relief, a Relief of 100m could be called Gentle or Moderate Relief. When Relief equals 100m, the membership of "Gentle Relief" is 0.5 as is the membership of "Moderate Relief".

4. Knowledge-Based Physiographic Analysis

In all earlier efforts in constructing prototype expert terrain related systems, knowledge related to the physiographic region of a site was not explicitly represented and used. It is however evident that the photointerpreter, in deciding the landform of a site is studying first, among other things, the physiography of a region and performs a kind of physiographic analysis and reasoning in order to create reasonable hypotheses of the possible landforms of the site [8, 10, 13, 14, 17, 19]. This type of reasoning is termed here physiographic context reasoning. On the other hand if the photointerpreter has already identified a landform, then she or he is in a position to create physiographic region hypotheses and, consequently, to be guided to interpret additional landforms on the basis of their spatial associations to the already interpreted landforms. Physiographic context and spatial reasoning are informal tasks at present, since they are not described explicitly in a formal manner in books and guides.

Argialas and Miliaresis [4, 5, 6] developed a formal conceptual framework for the representation of physiographic and spatial context reasoning within an expert system. Emphasis was placed on the definition of sub-problems and sub-tasks through domain-dependent concepts, hypotheses and observations. The presented case study concerns typical terrain of the Basin and Range Province of Southwest USA. The knowledge representation encompasses the typical physiographic sections of the Basin and Range province (Great Basin and Sonoran Desert) and the typical landforms of the piedmont slope and basin floor. The developed conceptual schemes are being used in the Terrain Analysis eXpert -4 (TAX-4) and -5 (TAX-5) systems which (a) guide the user in establishing tentative hypotheses about the types of physiographic regions based on observed evidence and (b) suggest reasonable landform hypotheses to be investigated. The role of landform spatial knowledge was developed both in helping the user to identify logical neighbours of an interpreted landform and to test the adjacency relations among interpreted landforms. In the design of TAX-4 and -5 the Smart Elements expert system tool (of Neuron Data) was used [6].

Detailed, "book-level" knowledge pertaining to physiographic regions and landform spatial associations was identified, named, described and organized. For the structural representation of physiographic and spatial reasoning knowledge an object-oriented representation structure was introduced. This uses frames as classes, subclasses, objects, and sub-objects, and slot frames as properties. The following terrain classes were named and described by their properties: physiographic regions,the Basin and Range concept, the Basin and Range youthful stage concept, the Basin and Range maturity erosion stage, plus a number of others.

Physiographic, topographic, and landform classes were organized into class-subclass hierarchies.Classes included sub-classes so that additional levels of detail were described only in the subclasses. Describing classes through subclasses gave access to a hierarchical representation of concepts and objects. Figure 4.1 shows the landform class-subclass hierarchy where the landform class (Landform Top) is the root under which are linked the subclasses containing various aspects of landforms: landform pattern elements (LF_PE), spatial reasoning indicators (LF_SR), engineering property indicators (LF_Engineering), suitability indicators (LF_Suitability), and military suitability indicators (LF_Military). Figure 4.2 shows the physiographic province hierarchy. The root superclass name is Physiographic Provinces and has as subclasses all the provinces. The subclass Basin and Range is linked to this superclass. The Basin and Range Youthful Stage and Basin and Range Maturity Stage are subclasses of the Basin and Range class.

While classes were useful in representing a concept as a whole, it was necessary to define individual (static or dynamic) class members or object instances of each class or subclass in order to use them for symbols for the interpreted features of each class on an image. Figure 4.3 shows a variety of class-instances for the classes of landforms, topographic forms, and physiographic regions.

Each class was defined by a set of properties indicating their distinguishing characteristics. Objects and subclasses can obtain their properties dynamically from a particular class through inheritance. Thus, through the class-subclass or class-instance hierarchy, these properties are inherited down through each hierarchy so as to be shared by all the members and instances of each class (figure 4.3). An object-sub-object or whole-part hierarchy was also defined in order to capture the whole-part terrain organization. In particular, a physiographic region (PH-1) was partitioned to its component physiographic features (PF-1), a physiographic feature to its component topographic forms (TF-3), and a topographic form to its component landforms (LF-1, LF-2) (figure 4.3).

A rule-base was developed for representing the strategic knowledge needed for inferring a physiographic region (province and section) from its own indicators [4, 5, 6]. For the Basin and Range concept, refinement rules inferred the concept of either a youthful or mature erosion stage, the first correspond-

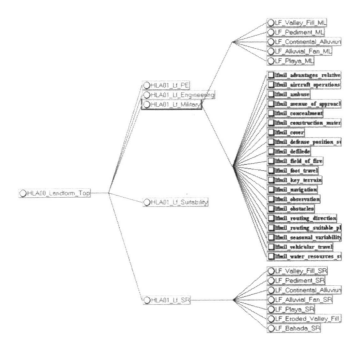

Fig. 4.1. The class-subclass landform hierarchy [6].

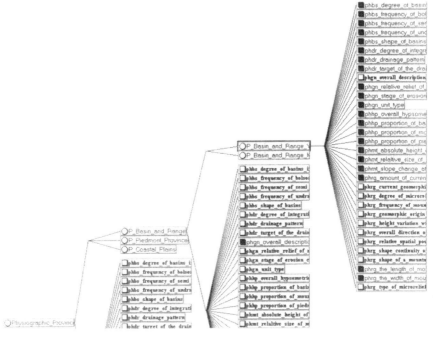

Fig. 4.2. The class-subclass hierarchy of physiographic provinces [6].

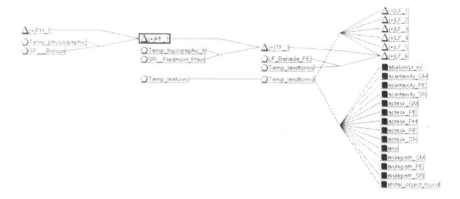

Fig. 4.3. Object-sub-object and class-instance relations [6].

ing to the Great Basin and the second to the Sonoran Desert section. One of the rules that infers the Maturity Erosion Stage concept follows.

IF relative_relief_of_region is "low"
and relative_size_of_mountains is "small"
and slope_change_at_piedmont_angle is "not abrupt"
and shape_of_basins is "rather plain than concave"
and overall_hypsometric_distribution_within_the_section is "more than 1/2 of the surface is below 2000 ft"
and proportion_of_Mountain_Ranges versus Piedmont_Plains versus Basins is "20% : 40% : 40%"
and amount_of_observed_tectonic_evidences_in_mountain_ranges is "low (the minority has a fault origin)"
and degree_of_basin_integration is "high"
and stage_of_erosion_cycle is "maturity (advanced,late)"
and frequency_of_bolsons is "low (less prelevant)"
and frequency_of_semi_bolsons is "high (more prelevant)"
and degree_of_integration_of_drainage_pattern is "high"
and outlet_of_the_drainage_network is "usually to another drainage basin"
THEN Basin_and_Range_Maturity_Stage is true and certainty is medium.

A rule-base was also developed for representing the strategic knowledge needed for spatial reasoning which included three distinct aspects a) landform identification by spatial association, b) landform verification by spatial association, and c) landform hypotheses-formulation by spatial association [4, 5, 6]. The landform identification by spatial association was developed in order to identify a landform by using its relevant spatial indicators (pattern elements). The landform verification by spatial association was developed in

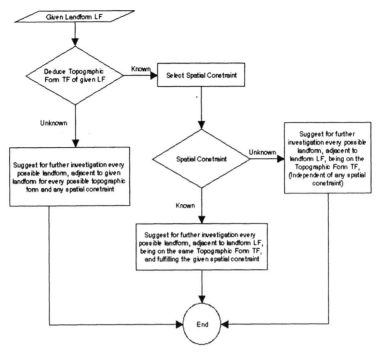

Fig. 4.4. Landform Hypotheses Formulation by Spatial Association [5].

order to test if two or more landforms, identified by the pattern element approach, were satisfying the required regional spatial constraints, as these were determined by geomorphologic and physiographic considerations. The landform hypotheses-formulation by spatial association, was developed so that once a landform was identified by pattern elements, the landform spatial knowledge rule-base suggested a small set of candidate landform hypotheses to be investigated by the user as being the most promising neighbouring landforms according to geomorphologic constraints. Figure 4.4 shows the reasoning behind the Landform Hypotheses Formulation by Spatial Association.

5. The Terrain Visual Vocabulary (TVV)

The TAX knowledge bases described earlier use many terrain indicators, each of which may have multiple values (e.g., the drainage pattern could be one of at least thirty types). For the most rudimentary knowledge base, the user must be familiar with more than three hundred different terrain indicators. Many more terrain features are needed for knowledge bases of significant size and detail.

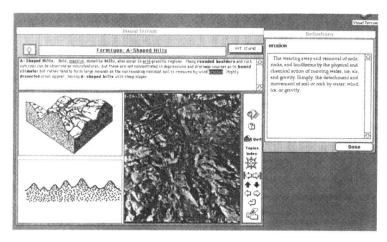

Fig. 5.1. A typical card from the Hypermedia Terrain Visual Vocabulary presenting the landform A-shaped Hills [5].

A novice user is unlikely to have unambiguous mental and visual models of all these terrain indicators. Accordingly, an interactive computer system could provide a terrain visual vocabulary (lexicon, or encyclopedia), which could incorporate explicit definitions, diagrams and photographic illustrations of each terrain indicator for use by novice users of expert systems like TAX. The prototype Terrain Visual Vocabulary (TVV) system was designed to include three components to define and graphically depict landform features: (1) definitions, (2) diagrams, and (3) aerial images.

The system was built in Hypercard, the authoring software environment of the Apple Macintosh computer. In the TVV, each landform indicator value was considered as a node and was represented by a card (figure 5.1). The set of all cards describing this prototype vocabulary was stored in one Hyper-Card stack. Each card contains a definition of the corresponding term in a text field, diagrams (profile and block diagram) of the vocabulary term, and aerial photographs exemplifying landforms for each term. The links between associated ideas (terrain terms) were implemented by constructing buttons or anchors, shown as boldface or underlined print in the definition text field, and then associating a certain Hypertalk language script to each of them.

6. Physiographic Feature Extraction, Representation, and Classification from Moderate Resolution Digital Elevation Models

The knowledge-based representation of the physiographic analysis process for the Basin and Range Province, presented earlier, has indicated the need for the extraction and classification of the common physiographic features such

82   D. Argialas

as mountain ranges, basins, and piedmont slopes from digital elevation models and/or satellite images. To address this need, Miliaresis and Argialas [11, 12] developed a methodology (a) for the classification of the GTOPO30 digital elevation model [18] to three classes of physiographic features (Mountains, Piedmont Slopes and Basins) and (b) for the identification, representation, and classification of mountain objects. The terrain classification methodology was applied to the Great Basin region of south-west U.S.A. where more than one hundred narrow mountain ranges separated by almost level desert basins have been observed [8].

After the rectification and resampling of the GTOPO30 digital elevation model, the Z operator, adjusted in a 9×9 neighbourhood, was first applied so that the elevation values of the DEM points were averaged over a 3×3 neighbourhood and then the partial derivatives were determined by the Sobel operator and the Z operator. Finally, the gradient and aspect at each DEM point were computed. For the computation of region growing criteria, training areas were selected and gradient statistics were computed and analysed to yield the following criteria: a) gradient less than 3 degrees for basins and b) gradient greater than 6 degrees for mountains.

The runoff simulation technique was used for the selection of an initial set of mountain and basin points since DEM points with large runoff belong either to the drainage network (downslope flow) or to the ridge network (upslope flow). An iterative region growing segmentation algorithm [15] was applied to the initial set of mountain points so that if a point with gradient greater than 6 degrees, was an 8-connected neighbour to the set of mountain points then it was flagged as a mountain point. An 8-connected component labelling algorithm [15] was applied and the regions formed by 8-connected pixels of either mountain or non-mountain points were identified. Then the analogous regions were recognised, on the basis of their size and spatial conditions, and they were reclassified. Finally, 36 distinct mountain objects were identified and a unique label identifier was assigned to every point belonging to a certain object (figure 6.1(a)). Similar region-growing procedures were applied for the classification of basin points. The points that were neither classified as mountain points nor as basin points were assigned to the piedmont slope class.

The boundary of each one of the 36 distinct mountain objects was delineated and skeletonised [5] to a one pixel wide line segment (figure 6.1(b)). A set of attributes were defined so as to carry sufficient physiographic information in order to be useful for the classification of mountain objects. The attributes included the area of the region occupied by the object, the object diameter, the mean polar radius, the eccentricity, the compactness, the polar index, the orientation, the asymmetry in orientation, the elevation, the roughness, the local relief, the relative massiveness, the elevation uniformity, and the slope. These attributes were used as descriptors for the parametric representation of mountain objects.

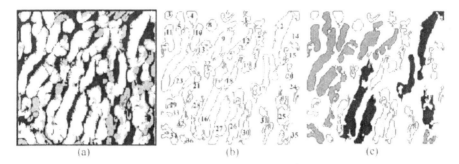

Fig. 6.1. (a) DEM classification to mountain (white), piedmont slope (grey), and basin (black) points [11], (b) Identified and labelled mountain objects, (c) Spatial distribution of the four classes of mountain objects [12].

The classification of mountain objects was achieved through the implementation of a K-means clustering algorithm applied to the parametric representation of mountain objects. The mountain object clusters were analysed and interpreted in terms of their characteristic properties, e.g. size, slope, elevation, relief, elongation, and orientation. The members of each cluster were depicted with a unique shade of grey in order to detect any spatial arrangement of the cluster members (figure 6.1(c)). It was observed that the mountain objects were grouped into four clusters that appeared to be spatially arranged to distinct geographic regions.

For evaluation purposes, the boundaries of the extracted mountain objects were delineated and superimposed on the digital elevation model of the study area and it was observed that they enclosed the elevated features observed on the model. The detected mountain features were also compared with the physiographic map of Atwood [7] and a good match was evident. The parametric representation of mountain objects, on the basis of the proposed geomorphometric attributes, and their classification into four clusters, spatially arranged to distinct geographic regions, was found to be in accordance with the existing physiographic descriptions and physiographic maps of the study area [7, 8].

7. Conclusions and Prospects

The structural drainage-pattern recognition system (DPA-PC) successfully recognized eight drainage pattern types. It needs to be extended to include more types of drainage patterns, to be integrated with a network extraction system (DEM-to-watershed transformation), to incorporate 3-D geomorphometric attributes derived from a DEM, and to be integrated with the terrain-analysis expert-system for mutual support of both systems.

The Terrain Analysis Expert system prototypes (TAX-1 to 5) involved the identification, conceptualization, and representation of knowledge related to landforms, topographic forms, physiographic features, and physiographic regions. They succeeded in capturing a number of "intermediate-level concepts" which are perhaps the most important tools available for organizing knowledge bases, both conceptually and computationally. The TAX systems need to be expanded in order to represent additional auxiliary information, including existing maps, structural and tectonic geology, soil data (and maps), land cover or land use maps, and relevant world knowledge.

Major types of links or associations need to be identified and represented between terrain related objects within the terrain visual vocabulary (TVV) system. Indeed, good links could help to categorize terrain related concepts into semantically and perceptually related units which will be linked and accessed by association, much as a human does. The challenge in choosing nodes and links is to structure the terrain related knowledge to reflect the mental models that experts create when they reason about terrain. The marriage of hypermedia and expert systems is also inevitable as hypermedia systems can help to build less ambiguous decision support systems and knowledge-bases and thus make expert interpretation systems more intuitive.

The geomorphometric processing of the digital elevation models to extract physiographic features and represent and classify mountain objects was successful as was indicated by the evaluation process. Integrating geomorphometric, spectral, and semantic aspects of terrain analysis problem solving could lead to significantly improved systems.

References

1. D. Argialas and C. Harlow, "Computational image interpretation models: an overview and a perspective", *Photogrammetric Engineering and Remote Sensing*, vol. 56, no. 6, pp. 871–886, 1990.
2. D. Argialas, "Towards structured knowledge models for landform representation", *Zeitschrift fur Geomorphologie*, N.F Suppl.-Bd. 101, pp. 85–108, Berlin-Stuttgart, 1995.
3. D. Argialas and E. Roussos, "DPA-PC: A drainage pattern recognition system in the PC environment", *Second National Conference of the Hellenic Committee for the Management of Water Resources: Integrated Approaches for Flood Danger Reduction*, pp. 165–174, January 12–13, 1995 (in Greek).
4. D. Argialas and G. Miliaresis, "Physiographic knowledge acquisition: identification, conceptualisation and representation", *Annual Convention of the American Society for Photogrammetry and Remote Sensing*, vol. 1, pp. 311–320, 1996.
5. D. Argialas and G. Miliaresis, "Landform spatial knowledge identification, conceptualisation, and representation", *Annual Convention of the American Society for Photogrammetry and Remote Sensing*, vol. 3, pp. 733–740, 1997.
6. D. Argialas and G. Miliaresis, "An object oriented representation model for the landforms of an arid climate intermontane basin: case study of Death Valley, California", *23rd Annual Conference of the Remote Sensing Society*, pp. 199–205, 1997.

7. W. Atwood, *Physiographic Provinces of North America*, Boston Massachusetts: Ginn & Co., 1965.
8. N. Fenneman, *Physiography of Western United States*, New York: McGraw-Hill, 1931.
9. A. Howard, "Drainage analysis in geologic interpretation: a summation", *American Association of Petroleum Geologists*, vol. 55, pp. 2246–2259, 1967.
10. T. Lillesand and R. Kiefer, *Remote sensing and image interpretation*, 3rd edition, New York: John Wiley & Sons, 1994.
11. G. Miliaresis and D. Argialas, "Physiographic feature extraction from moderate resolution digital elevation data", *24th Annual Conference of the Remote Sensing Society, RSS98, "Developing International Connections"*, 9–11 September 1998, The University of Greenwitch, pp. 545–551, 1998.
12. G. Miliaresis and D. Argialas, "Parametric representation and classification of mountain objects extracted from moderate resolution digital elevation data", *Proceedings of the fourth Annual Conference of the International Association for Mathematical Geology, IAMG98*, A. Buccianti, G. Nardi and R. Potenza (eds.), Isola d'Ischia, Naples, pp. 892–897, October 1998.
13. O. Mintzer, "Research in terrain knowledge representation for image interpretation and terrain analysis", *U.S. Army Symposium on Artificial Intelligence Research for Exploitation of Battlefield Environment*, El Paso, Texas, pp. 277–293, Nov. 1–16, 1988.
14. C. Mitchell, *Terrain evaluation*, London: Longman, 1973.
15. I. Pitas, *Digital image processing algorithms*, London: Prentice Hall, 1993.
16. A. Rosenfeld, and A. Kak, *Digital picture processing*, Florida: Academic Press, 1982.
17. J. Townshend (ed.), *Terrain analysis and remote sensing*, London: Allen and Unwin, 1981.
18. U.S. Geological Survey, "GTOPO30: 30 arc Seconds Global Digital Elevation Model", http://edcwww.cr.usgs.gov/landdaac/gtopo30/gtopo30.html, 1997.
19. D. Way, *Terrain analysis*, New York: McGraw-Hill, 1978.

Part II

High Resolution Data

Environmental Mapping Based on High Resolution Remote Sensing Data*

Kolbeinn Arnason and Jon Atli Benediktsson

Engineering Research Institute, University of Iceland,
Hjardarhaga 2–6, 107 Reykjavik, Iceland.
e-mail: {karnason,benedikt}@hi.is

Summary. Iceland is a sparsely inhabited country with important natural resources in the isolated areas. During the last two decades, remote sensing has been used both for monitoring natural resources and for environmental mapping in Iceland. For some purposes it has been noted that second generation satellite sensors such as the Landsat Thematic Mapper (TM) have not generated data with high enough spatial resolution. Therefore, high resolution airborne sensors have been designed at the Engineering Research Institute of the University of Iceland in Reykjavik, to collect data with the necessary spatial resolution. These specially designed sensors have spatial resolution of only a few meters compared to the 30m resolution of Landsat TM. In this chapter, the difference between the high resolution data and satellite data is discussed. Also, the need for the high resolution data is justified. Results of experimental studies using the high resolution data are shown. These studies consist of soil erosion mapping. Soil erosion is considered to be the most severe environmental problem in Iceland. The methods applied for soil erosion mapping are based on image processing and computer vision approaches such as non-linear filtering. With the use of the high resolution data not only the current status of the soil erosion but also the spatial distribution of the erosion activity can be mapped and classified adequately.

1. Introduction

Iceland is a 103.000 km^2 island in the North Atlantic Ocean, between latitudes 63° and 66°. The climate on the island is considered humid cold temperate to low arctic; volcanic eruptions are frequent and volcanic ash deposits are widespread. Iceland was settled 1100 years ago and is currently inhabited by about 270.000 people. After the settlement, rapid population growth led to intense use of a fragile ecosystem that had evolved in the absence of grazing animals. Degradation was severe and erosion accelerated significantly. The soil cover in Iceland is discontinuous and delicate. The interior is more or less barren [1, 7].

Soil erosion in Icelandic rangelands is considered the most severe environmental problem in the country. To attack this problem and combat the desertification processes systematically, it is necessary to accurately map the areas of soil erosion activity. For this purpose, remote sensing should be a desirable choice. However, the spatial resolution of the sensors on the current

---

* This research is supported in part by the Icelandic Research Council and the Research Fund of the University of Iceland.

remote sensing satellites, such as Landsat and SPOT, is not high enough to detect important surface detail necessary for many mapping purposes, including the mapping of soil erosion activity. To overcome this problem, researchers at the Engineering Research Institute of the University of Iceland have designed their own high resolution multispectral sensors.

In this chapter, the mapping of soil erosion using high resolution multispectral data will be discussed. Section 2 discusses soil erosion in Iceland. The applied imaging equipment is described in section 3. Image analysis methods used in experiments are reviewed in section 4. Experimental results are discussed in section 5, and finally, conclusions are drawn in section 6.

2. Soil Erosion in Iceland - Erosion Escarpments

Iceland is one of the most active volcanic areas on Earth. Therefore, Icelandic soils are rich in volcanic ash (tephra) which results in unique properties, some of which are responsible for their erosion susceptibility. The soils have low cohesion and wind erosion is intensified by the low density of soil grains, especially coarse ash grains (often about $1g/cm^3$). Soil erosion in Iceland is characterized by a total removal of the soil resource. Rich soil surfaces are denuded which leaves behind infertile barren surfaces. Soil erosion in Iceland has been divided into several categories based on classification of erosion forms that can be identified on the landscape. One of the major classes of erosion form consists of erosion escarpments which are ledge like phenomena in areas of thick (>30 cm) soil that is usually rich in tephra. In areas where erosion escarpments prevail the surface consists essentially of two classes: patches of vegetated ledges on a more or less barren background [1].

Figure 2.1 shows the distribution of erosion escarpments classified according to erosion activity. Areas of erosion escarpments are mainly confined to the active volcanic zone and its flanking areas which cross the island from the south-west to the north-east.

A photograph of a typical erosion escarpment is shown in figure 2.2. The soil is gradually eroded away by wind and horizontal rain leaving the land barren and infertile. As the erosion proceeds some of the wind blown soil grains from the escarpments are deposited on the vegetated ledge causing a gradual soil thickening and making the ledge still more vulnerable to the erosion process. When erosion for some reason ceases, these escarpments gradually level and disappear as vegetation slowly spreads from the ledges and the barren land is recovered. Therefore, the abruptness of individual escarpments is a measure of the erosion activity in the region of concern.

A schematic drawing of erosion escarpments in various stages of erosion activity (horizontal view) is shown in figure 2.3. Eroded barren ground is on the left and the vegetated ledge to the right. With processing of high resolution image data it is possible to map and classify the abrupt transition between full vegetation cover and barren land in areas of erosion escarpments.

Environmental Mapping Based on High Resolution Remote Sensing Data

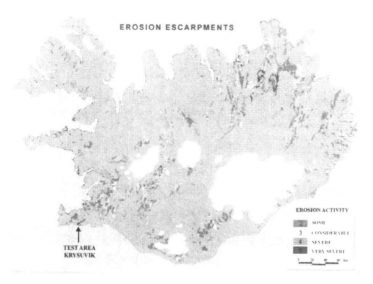

Fig. 2.1. The distribution of erosion escarpments classified according to erosion activity The figure also shows the location of Krysuvik which is used in experiments in section 5 [1].

Fig. 2.2. A typical erosion escarpment. The soil is gradually eroded away by wind and horizontal rain leaving the land barren and infertile. The dominant wind direction is from the right to the left [1].

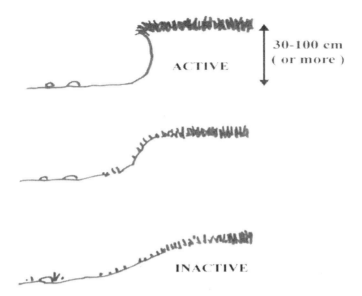

Fig. 2.3. Schematic drawing of erosion escarpments at three stages of erosion activity (horizontal view). Top: Active stage. Middle: Intermediate stage. Bottom: Inactive stage. Eroded barren ground is on the left and the vegetated ledge to the right.

3. The Imaging Equipment

Researchers at the Engineering Research Institute of the University of Iceland have designed their own image equipment to collect high resolution multispectral remote sensing imagery. This imaging equipment consists of an array of four identical CCD cameras sensitive in the visible and near infrared (0.4 − 1.0μm). Exchangeable optical band pass filters can be put in front of the lenses to provide for multispectral image acquisition. Green, red and infrared filters corresponding to the Landsat TM bands 2, 3 and 4 are mainly used for vegetation mapping purposes on three of the cameras. The fourth camera is often run without a filter (panchromatic mode). The optical axes of the cameras are parallel so at any instance they detect the same ground frame. The cameras are synchronized and the simultaneously acquired image data from all cameras is stored in an on-board computer. Minor differences between the geometry of image frames from individual cameras are corrected with postprocessing of the data. The imaging rate is adjusted with regard to altitude and ground speed of the aircraft to provide desired overlap between neighbouring image frames. Coverage of large areas is achieved by building mosaics of individual images. The current cameras have a 512 × 380 pixel array and the spatial resolution is only a function of the lens focal length and the flight altitude. The data discussed in this paper were acquired from an altitude of 8000 feet with a spatial resolution of 4.4m.

4. Applied Analysis Methods

4.1 Non-Linear Filtering

Although linear filters are easy to analyze, they have the major problem of deforming the signal which is being filtered. Therefore, non-linear filters have been designed to handle specific tasks and to overcome some of the problems associated with linear filters [3]. There are many different types of non-linear filters such as rank order filters (e.g. the median filter), morphological filters, and nonlinear means filters (figure 4.1). It is possible to consider traditional linear filters as special cases of non-linear filters.

Non-linear filters can be defined by:

$$y = f\left(\frac{\sum_{i=1}^{N} a_i \, g_{(i)}(x_i)}{\sum_{i=1}^{N} a_i}\right) \quad (4.1)$$

where:

- g and f are memoryless functions.
- a_i are the coefficients of the filter.
- x_i $i = 1, ..., N$ are the N pixel values that fall into a specific window.
- $g_{(i)}(x_i)$ are the N pixel values ordered according to a rank.

Many well-known filters can be grouped as non-linear filters. For example, if the functions g and f are defined as $g(x) = f(x) = x$ then a rank order filter is obtained. Also, the range edge detector [3] is obtained if the coefficients are selected as:

$$a_i = \begin{cases} -1 & i = 1 \\ 1 & i = N \\ 0 & i = 2, ..., N-1 \end{cases}$$

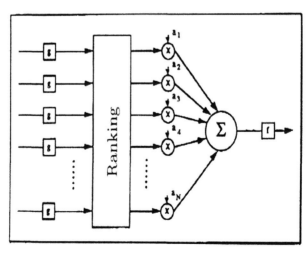

Fig. 4.1. A schematic of a non-linear filter.

94 K. Arnason and J. A. Benediktsson

By selecting the coefficients:

$$a_i = \begin{cases} -1 & i = 1, 2, ..., \left\lceil \frac{N}{2} \right\rceil \\ 1 & i = \left\lfloor \frac{N+1}{2} \right\rfloor, \left\lfloor \frac{N+1}{2} \right\rfloor + 1, ..., N \end{cases}$$

the Dispension Edge Detector (DED) filter is obtained [4, 5]. Both the range edge detector and the DED filter are edge detectors.

Morphological filters can also be designed using the non-linear filter coefficients above. For example, dilation [3] is obtained by selecting:

$$a_i = \begin{cases} 1 & i = N \\ 0 & i = 1, ..., N - 1 \end{cases}$$

and erosion by:

$$a_i = \begin{cases} 1 & i = 1 \\ 0 & i = 2, ..., N \end{cases}$$

In experiments we have concentrated on the use of the DED filter. DED is an edge detector which gives a strong signal at edges although the signal may be very noisy. For this reason, the DED filter is very well suited for detection of erosion edges in high resolution remote sensing imagery.

4.2 Classification

For classification the simple Normalized Difference Vegetation Index ($NDVI$) can be used. It can be calculated as follows:

$$NDVI = \frac{IR - VIS}{IR + VIS} \tag{4.2}$$

where IR is the grey level value of an infrared band for a pixel and VIS is the grey level value of the corresponding pixel in a visible band (red). The idea is that the greater the value of the IR return compared to the visible, the larger the vegetative index. The sum in the denominator is used to normalize the quantity. The $NDVI$ provides a relative indication of the vegetation, rather than an absolute one. It is also dependent upon the degree to which the relative size of the IR band response to that in the visible implies vegetation and only vegetation [6]. Rocks and soils in Iceland are typically very dark (of basaltic composition) and have low reflectance throughout the visible and near infrared portion of the electromagnetic spectrum. Therefore, they are easy to discriminate from vegetation. In our experiments the $NDVI$ was used to classify an image from an area of active soil erosion into classes of relative degree of vegetation cover.

5. Experimental Results: Mapping of Soil Erosion in Krysuvik

The test area, Krysuvik, is located in south-west Iceland and is marked on figure 2.1. This area is characterized by severe degradation of the soil escarpment type [1, 2]. In figure 5.1 two images of different spatial resolution from this area are shown. The image on the left is band 4 of a Landsat TM image. The image on the right is a high resolution aerial image obtained by the sensor designed at the University of Iceland. Both images were acquired in the same near infrared spectral band and differ only in spatial resolution. The Landsat TM image has a pixel size of 30 m vs. a pixel size of 4.4 m for the high resolution image. The images show the same 3 km × 2.2 km scene at the western border of the erosion area in Krysuvik. At the bottom of the images, part of the southern shore of Iceland is visible, and to the left there is a lava flow from the 12th century with a relatively homogeneous vegetation cover. In other parts of the scene, intensive soil erosion (of the erosion escarpment type) prevails. On these near infrared images, vegetation is light grey, barren ground is dark grey, and water is black. By looking at the images, it is obvious that the spatial resolution of the Landsat TM is not sufficient for studying soil erosion of this type in Iceland.

Another near infrared image from the Krysuvik test area is shown in figure 5.2. The image covers 6.6 km^2 (3 km × 2.2 km) which is about 20% of the whole erosion area at Krysuvik. This image was used in experiments

Fig. 5.1. Band 4 of a Landsat TM image from the test area (left) and the corresponding wavelength band of a high resolution aerial image (right). Both images show the same scene and differ only in spatial resolution.

Fig. 5.2. Near infrared high resolution image from the test area.

for testing several non-linear filtering algorithms as explained in section 4. Of those, the DED filter gave the best performance. As in figure 5.1 vegetation cover is light grey, barren ground is dark grey and water is black. The image in figure 5.2 has been rotated so that north is to the left and the shoreline is to the right. Note that there are both abrupt and gradual changes from barren to vegetated areas (from dark grey to light grey) on the image.

A classification of the scene in figure 5.2 is shown in figure 5.3. In the classification process four vegetation cover classes and the water class were determined by thresholding the NDVI image. The upper pie chart in figure 5.3 shows the proportional sizes of the four land cover classes when water is excluded. The two extreme classes, barren areas and 100% covered areas, are most prominent, the two intermediate classes being subordinate and only visible where there is a gradual transition from fully vegetated to barren land.

The image in figure 5.4 shows the result of filtering figure 5.2 using the DED filter with a 4 by 4 window. The more abrupt the change is from vegetated to barren land (or the greater the changes in grey values between adjacent pixels of the original image) the higher the resulting grey value in the filtered image. The highest grey values occur where the two classes, barren land and 100% vegetation cover, are adjacent to each other. Gradual change between these two classes gives lower grey values and no change between neighbouring pixels results in a zero pixel value (black).

Classified results of the filtered image in figure 5.4 are shown in figure 5.5. The classification was done by thresholding the grey values of figure 5.4 and thereby dividing the grey scale into 4 classes where class 0 represents no (or very little) change between neighbouring pixels in figure 5.1, class 3 represents

Environmental Mapping Based on High Resolution Remote Sensing Data 97

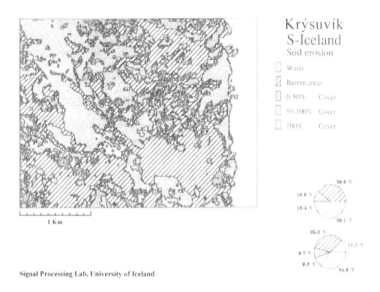

Fig. 5.3. Classification map obtained by the use of NDVI for the scene in figure 5.2.

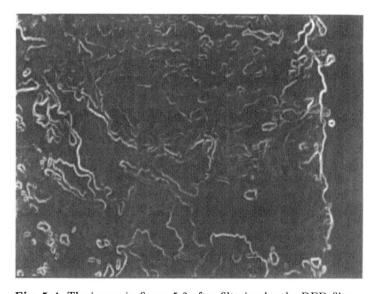

Fig. 5.4. The image in figure 5.2 after filtering by the DED filter.

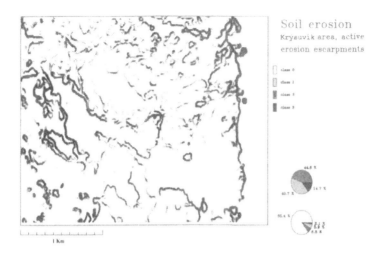

Fig. 5.5. Classification map obtained by the use of thresholding operation on the filtered image in figure 5.4.

an abrupt change from the lowest to the highest pixel values, and classes 1 and 2 are intermediate classes. This image is a black and white version of a colored original which makes it difficult to discriminate between class 2 (red) and class 3 (black). The three escarpments classes (class 0 is excluded) in figure 5.5 correspond roughly to figure 2.3 in such a way that the active erosion escarpment (top of figure 2.3) relates to class 3, the middle figure to class 2, and the inactive erosion escarpment (bottom of figure 2.3) to class 1. These results correspond very well to results in [2] which were obtained independently by the use of different approaches.

6. Conclusions

In remote sensing, spectral information is necessary for vegetation mapping and classification and high spatial resolution is essential for detecting and mapping fine textural features. Many natural phenomena in Iceland are of either fine texture or limited spatial extent. However, the spatial resolution of conventional satellite sensors (Landsat TM, SPOT) is not applicable for mapping and analyzing these phenomena. In this chapter, we have examined a problem where the combination of spectral information and high resolution spatial information is essential.

In our experiments, erosion structures (fronts or escarpments) were mapped using high resolution image data. These escarpments were classified into several erosion activity classes in an area of heavy land degradation. Using this methodology, it has been shown to be possible to map erosion over a wide area and to identify the locations of the currently most severe erosional activity. Such erosion activity maps derived from the structures present in classified high resolution remotely sensed imagery can be of vital importance when systematic revegetation of an area is planned.

References

1. O. Arnalds, E. F. Thorarinsdottir, S. Metusalemsson, A. Jonsson, E. Gretarsson, and A. Arnason, *Soil Erosion in Iceland* (in Icelandic), The Soil Conservation and Reclamation Service and The Agricultural Research Institute, Reykjavik, February 1997.
2. G. Gisladottir, "Environmental Characterisation and Change in South-western Iceland", Department of Physical Geography, Stockholm University, Dissertation Series, Dissertation no. 10, 1998.
3. A. K. Jain, *Fundamentals of Digital Image Processing*, Prentice Hall, Engelwood Cliffs, NJ, 1989.
4. I. Pitas and A. N. Venetsanopoulos, "Nonlinear Order Statistic Filters for Image Filtering and Edge Detection", *Signal Processing*, vol. 10, pp. 395–413, June 1986.
5. I. Pitas and A. N. Venetsanopoulos, "A General Nonlinear Filter for Image Processing", in *Proceedings IEEE International Conference on Circuits and Systems 1988*, pp. 949–952, IEEE, NJ, 1988.
6. A. J. Richardson and C. L. Wiegand, "Distinguishing Vegetation from Soil Background Information", *Photogrammetric Engineering and Remote Sensing*, vol. 43, no. 12, pp. 1541–1552, December 1977.
7. S. Thorarinsson, "Cohabitation of Land and People for Eleven Centuries" (in Icelandic), *Saga Islands I*, HIB, pp. 29–93, Reykjavik, 1974.

Potential Role of Very High Resolution Optical Satellite Image Pre-Processing for Product Extraction.

P. Boekaerts, V. Christopoulos, A. Munteanu, and J. Cornelis

Vrije Universiteit Brussel, Department of Electronics and Information Processing
Pleinlaan, 2 - B-1050 Brussels, Belgium.
e-mail: pkboekae@etro.vub.ac.be

Summary. Modern optical Very High Resolution (VHR) sensors boost the resolution of satellite imagery up to 1 pixel/m at nadir and higher. It is believed that the appearance of recognisable (man-made) structures and texture will drastically increase the number of data products and therefore also the number of end users. The potential role - and typical problems - of a selected set of image analysis tools for the pre-processing of VHR products is discussed.

1. Introduction

An important problem in the set-up of (pre-)processing chains for the commercial exploitation of optical satellite data is the identification of the end user and his processing needs.

Figure 1.1 shows that 1m resolution imagery, achievable by e.g. the Ikonos, QuickBird and future OrbView satellite series, reveals (man-made) structures and texture that cannot be recognised at more conventional resolutions (10m and lower). These two new features of 1m optical satellite data imply that the number and variety of data products that can potentially be derived from these data is much higher than for conventional resolution satellite imagery. Very High Resolution (VHR) data will not only improve the conventional objectives of existing satellite missions (e.g. SPOT, LANDSAT) with respect to e.g. cartographic applications, urban studies, monitoring of agricultural activity, ecological disasters, and forest, but will also stimulate new applications of remote sensing.

The potential role of image processing for the exploitation of VHR data is not only defined by the degree to which these new features - structure and texture - can be processed (semi-)automatically, but also by the level of processing that is considered.

At least four processing levels can be distinguished:

- sensor driven processing related to operations, including: calibration of the spectral channels, geometrical rectification, derivation of sensor specifications (e.g. spectral and spatial response functions)
- pre-processing for products, including: image enhancement, segmentation, edge detection, co-registration, data fusion

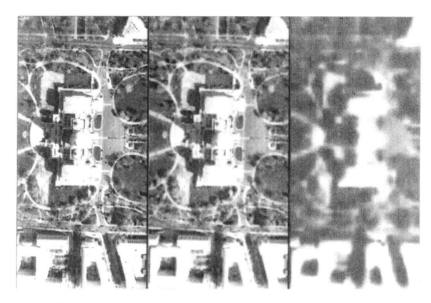

Fig. 1.1. Comparison of 10m (right), 3m (middle) and 1m (left) resolution imagery of the Capitol building, Washington DC. From "Space Connection 25, Telsat: dix ans de teledetection spatiale".

- product extraction, including: pattern recognition (e.g. morphological filtering), classification, tracking and change detection
- processing related to the archiving, transmission and consultation of data and products, including: data compression, progressive data transmission, data and product storage and representation (e.g. GIS)

The advent of very high resolution data (with appearance of new structure and texture) will primarily affect the usability of existing image processing tools at the pre-processing and product extraction levels.

The potential role of a selected set of existing image processing tools for the pre-processing of VHR products is discussed in the following sections. The reader is referred to existing literature for a technical review of the algorithms themselves.

2. Image Enhancement

The purpose of image enhancement is to improve the appearance of the image content to human viewers and/or to enhance the processing performance for product extraction. A wide set of standard image enhancement algorithms for contrast and/or dynamic range modification, noise reduction and/or edge enhancement exists, each one of which has a specific set of parameters to be tuned [6].

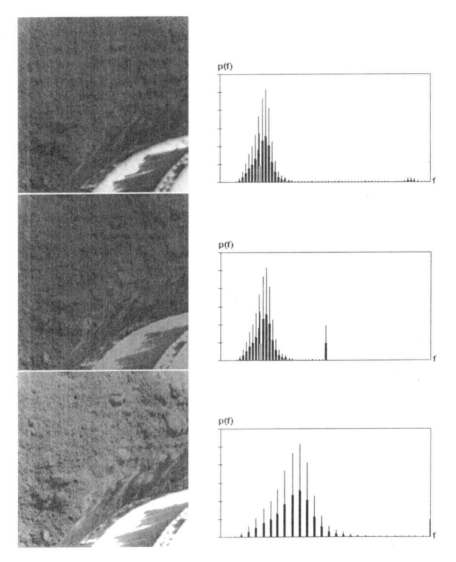

Fig. 2.1. Example of image enhancement. Top: original image and corresponding histogram. Middle: result obtained after clipping the histogram (clip value: 128). Bottom: image obtained after stretching the clipped image in the middle.

This is illustrated in figure 2.1 for the image enhancement technique known as "clipping before stretching". The small bright region in the right lower corner of the overall dark original image (top left), makes enhancement by simple histogram stretching impossible.

To increase the overall dynamic range of the image intensities in the dark region of the image, a clipping of the pixel grey values must first be applied, resulting in the middle left image. The clipping threshold (equal to 128 in this case) must be selected by examination of the histogram profile of the image (-region). The result of histogram stretching applied to the clipped image in the middle is shown in the bottom left, illustrating the appearance of texture that cannot (or hardly) be distinguished in the original image.

3. Image Segmentation

The purpose of image segmentation is to reduce the information content of an image without loosing structural information. Good segmentation algorithms should supply the user with a set of spatially coherent image regions that reflect structural information of interest. Segmentation algorithms that take into account the local spatial structure of images have been shown to guarantee the required spatial coherence of segmented image regions for structural analysis [2, 3, 4]. One of these segmentation algorithms consists of the vector quantisation (K-means) of the (complete) segment distribution of the image (covering all possible rectangular sub regions of the image, called image segments) [3] and is used for demonstration purposes in the examples of this section.

In what follows, we investigate two specific questions: does VHR imagery indeed contain redundant information for the analysis of structure, and, if so, do segmentation algorithms that take into account local structural information of the images indeed reduce the complexity of product extraction with respect to morphological filtering and object identification ?

For that purpose, different case studies have been carried out on 1m resolution airborne data of the Rotterdam harbour, including oil tank, ship, container, and vegetation monitoring. A reduced (grey value) overview of the original tri-spectral image source can be found in figure 3.1. The results of the two first case studies (oil tank and ship monitoring) are presented and discussed.

3.1 Remote Oil Tank Monitoring

An enlarged window of the bottom left corner of figure 3.1 representing an oil storage field is given in figure 3.2 (top left). The vertical height of the cover of an oil tank reflects the oil content of the tank. Full oil tanks are closed and the cover has in general a high visible reflectivity. Nearly empty oil tanks are

Fig. 3.1. Reduced global view of an airborne image of part of the Rotterdam harbour (1m resolution).

open and the reflectivity of the cover is reduced and spatially distorted by the shadow of the tank.

After applying a vector quantisation of the segment distribution of the original image with segments of 3 × 3 pixels [3] a segmented image with (in this case) only 20 grey values is obtained (top right). The typical features of full and nearly empty oil tanks are still visible for a human observer, which indicates that the relevant information for oil tank monitoring is preserved after reducing the information content of the image.

To see how far the segmentation result reduces the complexity of product extraction, the labels of those image regions belonging to oil tank covers with high reflection properties were merged and given one bright label (bottom left image). The bright circular covers indicate full oil tanks and can easily be detected by (conventional) morphological filters. The problem of full oil tank detection is even further simplified after merging the background labels, giving them one dark label and applying a 13 × 13 median filter to reduce spurious noise introduced by other objects with high reflection properties (bottom right image).

It is worth noting that the 10 examples of full oil tanks that obey the a priori defined properties or "rules" of appearance and that are easily detectable in the original image by a human observer, can all be detected automatically after segmentation of the original image.

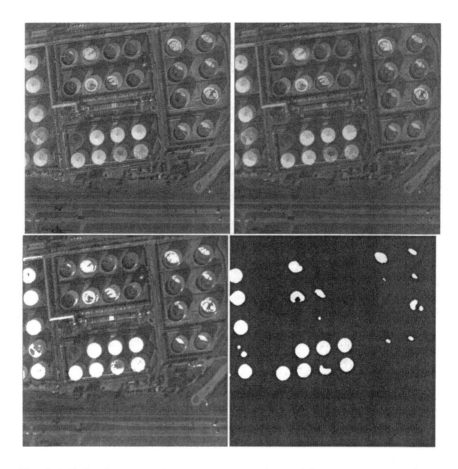

Fig. 3.2. Role of segmentation as a pre-processing tool for the detection of full oil tanks. Top left: original image; right: segmented image with 20 grey value labels. Bottom left: merging and re-labelling of regions belonging to the oil tank covers (white label); right: circular full oil tank regions obtained after removing other region labels and after applying a 13 × 13 median filter.

The final evaluation of the role of segmentation as a pre-processing step for oil tank monitoring depends however on the robustness of the assumptions made on the appearance of full oil tanks. The visible reflection properties of the covers and the amount of shadow in nearly empty oil tanks depends on the position of the sun for clear sky conditions and thus on the time of the day that the image was taken. The difference in possible illumination conditions certainly complicates the apparently simple problem of remote oil tank monitoring, but does not affect the usefulness of segmentation to model and solve the impact of illumination conditions on the detection of full oil tanks.

106 P. Boekaerts *et al.*

Of the 29 observable oil tanks, 3 can be distinguished however (i.e. the second oil tank from the bottom of the left single column of oil tanks, first oil tank of the first row and third oil tank from the second row of the bottom cluster of 2×4 oil tanks) that do not obey the assumed appearance rules and that are probably full oil tanks too. The reflection properties of the covers of these oil tanks are different (not homogeneously bright or even not at all bright in the second case) which makes these tanks harder to detect as being full, even by human inspection.

A visual inspection of the two oil tanks in the second column of the 2×4 oil tank cluster at the top suggests that these tanks are nearly full, which could only be detected in the filtered image (bottom right) by more complicated morphological filters (the same is true for 2 half full oil tanks in the 4×2 cluster of tanks at the right).

3.2 Remote Ship Monitoring

The same procedure has been applied to another sub region of figure 3.1 to analyse the potential role of segmentation for remote ship detection and ship activity monitoring.

Figure 3.3 (top left) shows the original image with a ship moored in a dock, surrounded by buildings (on land). The top right segmented image, obtained after vector quantisation of the segment distribution defined by segments of 3×3 pixels [3], confirms again that relevant structural information is preserved after reducing the information content of the image (to 20 labels).

An interference between different objects in the segmented image can be observed however. The water in the dock has a uniform (grey) label (with a small amount of speckle noise) that is also assigned to some vegetation structures and shadow. Image regions with high reflection properties on the ship deck do have the same label as parts of the roofs of the buildings.

In the bottom left image, shadow regions from the segmented image are given a black label and the bright reflection regions of the ship deck have been enhanced by a white label confirming the interference with part of the roofs of the surrounding buildings. The bottom right image shows that median filtering (5×5 pixels) followed by region merging and re-labelling (up to 5 labels in total) reduces further the information content of the image significantly without affecting the possibility of detecting the ship in the dock.

The presence of shadow regions and the interference of image regions belonging to different structures after segmentation do not simplify the problem of ship detection and monitoring.

The shadow regions of the segmented image are spatially coherent image regions with one unique label however, making it possible to identify these regions and subject them to further pre-processing (e.g. image enhancement or interpolation to remove shadow, or, morphological analysis to characterise the structures causing the shadow).

Fig. 3.3. Role of segmentation as a pre-processing tool for the monitoring of ship activity. Top left: original image; right: segmented image with 20 grey value labels. Bottom left: image obtained after merging and re-labelling the shadow regions (black) and after enhancing the regions with high reflection (ship deck and roofs = white); right: image obtained after applying a 5 × 5 median filter to the left image and after reducing the number of labels to 5 (including shadow, water, reflection of the ship and roofs and background).

The discrimination of different structures with the same segmentation label (roof, ship deck) requires the development of advanced morphological filters. The complexity of these filters can drastically be reduced, however, by the co-registration and fusion of the segmented images with cartographic data.

Similar conclusions could be formulated for the monitoring of container activity in the harbour, indicating the usefulness of segmentation for the purpose of pre-processing for VHR data products.

The potential role of segmentation for vegetation monitoring could not be confirmed however for reasons that are explained in section 4.

4. Texture Enhancement

For objects with a high structural variability, texture analysis can be at least as important for identification purposes as morphological analysis.

Texture classification tools rely on the explicit analysis of typical spatial distributions of grey (or colour) values in a local region of the image. For this reason, information reduction by means of segmentation (section 3), cannot be advised as an adequate VHR pre-processing approach for texture detection and texture identification.

Multi-resolution wavelet analysis [7] of textured regions has long been considered in the literature as an important breakthrough for texture detection and identification although finally, it has been confirmed that multi-scale texture identification in wavelet space does not realise the promised breakthrough [5, 8, 9].

A case study was done on figure 3.1 (middle-left part of the image) for an image region covering a park with vegetation and trees (figure 4.1, left) to see how far multi-resolution wavelet analysis could contribute to the pre-processing of textured regions. This aim was to enhance the textural information of such regions rather than reduce the information content of these regions (which is done for segmentation and classification purposes).

The texture enhancement illustrated in figure 4.1 (middle) was obtained as follows. The RGB components of the original tri-spectral image (left) were decomposed on 6 successive scales using biorthogonal 9–7 filters [1]. Biorthogonal 9–7 filters were used because they are known to represent well the high frequency components of the image [1].

Fig. 4.1. Texture enhancement in the wavelet space. Left: original image (grey value representation). Middle: reconstructed image using the maximum wavelet coefficients of the RGB components. Right: reconstructed image using the minimum wavelet coefficients of the RGB components.

Textured areas and edges are represented by large (in absolute value) wavelet coefficients. If the maximum of the (absolute value) of the coefficients of the three RGB components is taken for each scale, we can compose a set of wavelet coefficients from which a new mono-spectral image can be reconstructed. This reconstructed image (figure 4.1, middle) reveals a texture enhancement which becomes even more appealing when compared to the right image, reconstructed from the set of (absolute) minimum wavelet coefficients of the RGB components.

It is not clear for the moment, however, how far this type of texture enhancement can be useful in pre-processing of VHR products related to texture. Even the benefits of this type of enhancement technique with respect to conventional image enhancement procedures (section 2) is not proven yet.

An important drawback of the method for the pre-processing of RGB images is that colour information of textured regions is lost, although this argument is not relevant if one wants to enhance texture information from combined visible and infrared VHR imagery.

5. Conclusion

The potential role of a selected set of image processing tools for the pre-processing of structure and texture in VHR optical image data has been discussed. Different case studies showed that, although such tools can reduce the complexity of computer aided product extraction for a large set of applications, they need the supervision of operators in order to tune their parameters, to select the image regions where they can be applied, and to decide for which purpose they can be applied. Despite the variability of illumination conditions due to the sun's position and the variety of potential data products that can be extracted from VHR satellite data, we believe that existing pre-processing tools can play an important role in the development of new, advanced processing algorithms related to the analysis of structure and texture that will be needed for final product extraction from remotely sensed imagery.

References

1. M. Antonini, M. Barlaud, P. Mathieu and I. Daubechies, "Image coding using wavelet transform", *IEEE Transactions on Image Processing*, vol. 1, pp. 205–220, 1993.
2. P. Boekaerts, E. Nyssen and J. Cornelis, "Autoadaptive monospectral cloud identification in METEOSAT satellite images", *European Symposium on Satellite Remote Sensing, Conference on Image and Signal Processing for Remote Sensing II*, EOS/SPIE vol. 2579, pp. 259–271, 1995.

110   P. Boekaerts *et al.*

3. P. Boekaerts, E. Nyssen and J. Cornelis, "Autoadaptive scene identification in multispectral satellite data", *5th Symposium on Remote Sensing: A Valuable Source of Information*, NATO AGARD/SPP CP-582, pp. 21.1–21.8, October 1996.
4. P. Boekaerts, E. Nyssen and J. Cornelis, "A comparative study of topological feature maps versus conventional clustering for (multi-spectral) scene identification in METEOSAT imagery", in *Neurocomputation in remote sensing image analysis*, I. Kanellopoulos, G. G. Wilkinson, F. Roli and J. Austin eds., pp. 231–241, Springer Verlag, Berlin, 1997.
5. T. Chang and C.-C. J. Kuo, "Texture analysis and classification with tree-structured wavelet transform", *IEEE Transactions on Image Processing*, vol. 2, no. 4, pp. 429–441, 1993.
6. FUSETUTOR: Multi-media Tutorial on Remote Sensing Image and Data Fusion, Western European Satellite Centre, Madrid.
7. S. Mallat, "A theory for multiresolution signal decomposition: the wavelet representation", *IEEE Pattern Analysis and Machine Intelligence*, vol. 11, pp. 674–693, 1989.
8. M. Unser, "Texture classification and segmentation using wavelet frames", *IEEE Transactions on Image Processing*, vol. 4, no. 11, pp. 1549–1560, 1995.
9. Y. M. Zhu and R. Goutte, "Analysis and comparison of space/spatial frequency and multiscale methods for feature segmentation", *Optical Engineering*, vol. 34, no. 1, pp. 269–282, 1995.

Forestry Applications of High Resolution Imagery

Tuomas Häme, Mikael Holm, Susanna Rautakorpi, and Eija Parmes

VTT Automation, P.O. Box 13002, FIN-02044 VTT, Finland.
e-mail: {Tuomas.Hame,Mikael.Holm,Susanna.Rautakorpi,Eija.Parmes}@vtt.fi

Summary. In this paper a possible chain for the processing and interpretation of high resolution imagery is presented. It starts from the basic geometric and radiometric processing of the images and ends at the numerical interpretation approaches. With a spatial resolution of few meters new problems in the image analysis process are encountered. First, many images have to be combined into mosaics. Second, traditional pixel by pixel approaches in the interpretation are not successful when the neighbouring pixels represent tree crowns and the shaded space between the crowns.

A test example is shown using airborne digital high resolution imagery. It was possible to create high quality mosaics automatically using more than a hundred images. The combination of textural information with spectral information in the image interpretation seemed to be a sound base for further research.

1. Introduction

A result of several user requirements surveys on the applicability of spaceborne remote sensing in forestry has been that the spatial resolution is a limiting factor in local scale applications. This limitation will disappear when the new high resolution satellite sensors emerge. (In this paper high resolution means a ground resolution of one to a few meters). High resolution data, do not only offer an opportunity to drastically improve the accuracy of the estimates of the traditional forest characteristics, but it also makes it possible to compute variables describing the structural diversity of the forests. In addition digital elevation models can be generated automatically.

However, numerical interpretation of satellite data with a resolution of some meters is complicated. Traditional pixel by pixel approaches in the interpretation are not successful when the neighbouring pixels represent tree crowns and the shaded space between the crowns. The improved spatial resolution of the images increases the variability of radiances within stands which makes stand separation more difficult. A pixel by pixel classification, without incorporation of any textural features, may lead to poorer classification results although the images are more detailed.

The textural features of forest areas should reflect variation of tree size, shape and the spatial arrangement of the trees, i.e. the spatial interaction of individual trees. Hay *et al.* [5] present a structural approach, in which the tree peaks are networked into triangles. These triangles are used as texture primitives. The area, the average spectral values and variances of these texture primitives are used as features in the maximum likelihood classification.

112   T. Häme *et al.*

Statistical approaches use the variance and mean values from a pixel neighbourhood or the GLCM-method (Grey Level Co-occurrence Matrix), to calculate entropy, homogeneity or contrast values [4]. These textural features can be used as input variables in the classification.

Additional problems, associated with high resolution data, are a rather small image size which may require image mosaicking in the operative image analysis, and the poor radiometric resolution and signal to noise ratio due to weak radiance from the small ground unit.

In this paper a possible chain for the processing and interpretation of high resolution imagery starting from the basic geometric and radiometric processing of the images and ending to the numerical interpretation approaches will be presented. The results are based on the analysis of airborne digital high resolution imagery.

2. Materials and Methods

2.1 Study Site

The study site is located in southern Finland. The size of the whole test area was 2×7 km^2. Airborne digital camera data was gathered of the area at a ground resolution of 0.5×0.5 m^2. The 500×500 m^2 sub-area (figure 2.1) selected for demonstrating the processing and interpretation chain comprises boreal forest, dominated by pine and spruce, as well as tillage. In the middle of the area there is a stand of younger forest, which can be seen due to its finer texture. The eastern part of the image is older forest having a coarser texture. In practical forestry, it is important to distinguish these two forest categories, because the silvicultural measures needed are different. Also the biomass of the older forest with a coarser texture is higher.

2.2 Image Analysis Chain

Figure 2.2 shows the two main phases of the image analysis process: image correction and image interpretation. In the image correction part we present an integrated approach, which is in a semi-operative phase. This approach is compared with a modular approach, which is in operative phase. Image interpretation is still rather experimental.

2.3 Image Correction

The purpose of the image correction phase is to compile one accurate image mosaic out of many small (about 1500 by 1000 in the digital camera case) images, and to calibrate this mosaic radiometrically in order to be able to do valuable interpretation in the later phase. Most of the methods can later be

Forestry Applications of High Resolution Imagery 113

Fig. 2.1. Image mosaic of six images. Red band of a normal colour image, pixel size one meter, total image area 500 × 500 m^2.

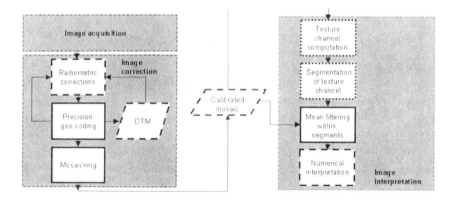

Fig. 2.2. Image analysis chain. Boxes with uniform border indicate elements that are already operative, whereas the dashed line boxes indicate semi-operative status of the method development and dotted lines indicate an experimental status.

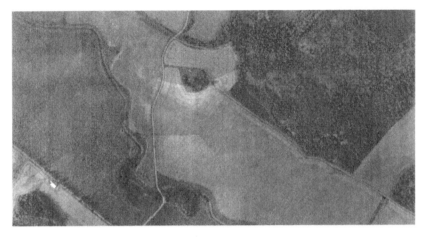

Fig. 2.3. A 15 image mosaic using bundle block adjustment (no DTM).

applied to high resolution satellite imagery, as off-nadir viewing leads to the same problem of height differences in the terrain.

The traditional modular approach is already available as an operative system. In this approach the first step in creating image mosaics out of many small images is to normalize the individual images with respect to the sun angle and the viewing angle, and with respect to sensor specific factors. Because the atmospheric model is normally not perfect and for instance the BDRF (bi-directional reflectance function) of all the different vegetation types in the terrain is normally not known there will remain radiometrical differences between the same object in the different images.

In the following phase the geo-coding of the imagery is done by means of a normal (semi-automatic) photogrammetric bundle block adjustment. After the bundle block adjustment, the position and attitude of each image is known, and also the projection from ground to image coordinates. Afterwards the mosaic is created in a batch process. The resulting mosaic (of another test site) can be seen in figure 2.3.

There are two visible drawbacks of the mosaic in figure 2.3. Firstly the seams between the images can be clearly seen due to radiometrical differences. These seams are caused by the fact that the images were captured during semi cloudy weather below the clouds. When the clouds moved, also their shadows moved. So, a sunny object could be in the cloud shadow in the neighbouring image. In other words, the atmospheric effects were not modelled correctly.

The second drawback is the high number of geometric errors in the mosaic, for instance along the creek and along the road. These errors are caused by height differences in the terrain and could be removed by using a Digital Terrain Model (DTM). In areas of a rough terrain a DTM is the only solution. The two dimensional (for instance polynomial) geo-coding approaches that are effective with lower resolution satellite imagery would not give good

Forestry Applications of High Resolution Imagery 115

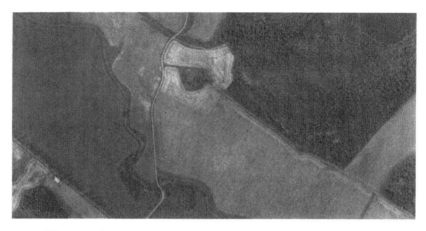

Fig. 2.4. A 15 image mosaic using global object reconstruction.

results when applied on high resolution imagery. Accurate DTM's are not available everywhere. However, it is possible to compute a DTM from the image data if the geo-coding is done accurately enough, in three dimensions.

The second and more advanced approach is called global object reconstruction or global matching. This integrated approach means development of a general model for digital photogrammetry, which integrates area-based multi-image matching, point determination, object surface reconstruction and orthoimage generation. Using this model, the unknown quantities are estimated directly from the pixel intensity values and from a control information in a nonlinear least squares adjustment. The unknown quantities are the geometric and radiometric parameters of the approximation of the object surface (e.g. the heights of a digital terrain model and the brightness values of each point on the surface), and the orientation parameters of the images. Any desired number of images, scanned in various spectral bands, can be processed simultaneously. In other words it integrates all the tasks in the image correction phase of figure 2.2. The result of the global object reconstruction method [6, 7] can be seen in figure 2.4.

In the mosaic of figure 2.4 there are no radiometrical or geometrical errors left and it is therefore much better suited for interpretation tasks. One reason for the seamlessness of the mosaic is that the computed DTM follows the highest dominating surface, for instance tree tops in dense forests and house roofs in towns. Therefore the orthoprojection of the dominating feature is done correctly from all images. In fact, as it is very difficult to do the physical modelling perfectly (e.g. the atmosphere and the BDRF) we always have to do the final adjustments based on the intensity values of each ground point projected to the images, and here we need to know the three dimensional relationship between the ground and the images. Therefore accurate DTM's are essential. And one big advantage of the method is that it is practically fully automated.

2.4 Image Interpretation

In the interpretation approach applied, a texture channel is first computed from the image. This channel should keep the borders of the target classes of the interpretation but should smooth out the irrelevant variation. We have used the standard deviation of the intensities as a textural channel. Tests showed that an appropriate size for the window within which the deviation is computed is 17m × 17m, i.e. 17 × 17 pixels in our image data using 1 × 1 m^2 resolution (figure 2.5). Such a window includes some 15 trees in a typical mature boreal forest. In a young forest the number of trees is four to five times as high within the same area.

In the second phase, the deviation channel is segmented. We used a segmentation algorithm by Narendra and Goldberg [9] with some improvements proposed in [10] (figure 2.6).

The third phase is computation of the intensity mean of each segment and filling the segments with the mean value. The segment-wise mean filtering was applied both to the original channels and to the deviation channels from the original bands.

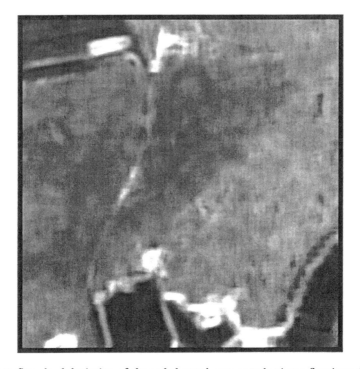

Fig. 2.5. Standard deviation of the red channel, computed using a floating window of 17 by 17 pixels.

Fig. 2.6. Segmentation of the deviation image.

The final phase is image classification. We have used an unsupervised classification method developed by Häme et al. [2]. In this method, a subsample is selected from the image, representing homogeneous ground targets. The clustering is performed to this sub-sample and as the last step each pixel of the image is classified using the spectral statistics of the clusters.

The clustering was applied to the three original channels, mean filtered by the image segments, to the three standard deviation channels, and to the combined six channel image, including both the original channels and the deviation channels. An option in the clustering procedure was selected, which normalizes the channels by dividing them by their standard deviation. Thus, each band had a similar weight to the clustering process. An adequate number for the clusters appeared to be fifteen for this area and data. The results were validated visually.

3. Results and Discussion

For the image correction the global object reconstruction approach has proven to be very useful. Although it still is in a pre-operational phase, mosaics have been created using hundreds of images. Figure 3.1 (left) shows a mosaic of the whole test site (2×7 km^2) consisting of 141 images. The geometry of the mosaic is visually excellent. The atmospheric corrections were not (yet) included in the process and therefore the radiometric stability of the mosaic is not completely satisfactory.

The classification results using either original channels or standard deviation bands had a similar general appearance but the result using the original bands was very fragmented. The best results were achieved in the clustering of the standard deviation bands only (figure 3.2). In this classification, the older forest was partially assigned to different clusters than the younger forest but the discrimination of these different age classes was not fully acceptable. The agricultural lands were well distinguished from the forests which is obvious. Also the small openings within the forest were assigned to their specific clusters. Shadows of the trees on the agricultural fields were classified as forests.

The utilized image data were natural colour images from the visible part of the spectrum. In forestry applications, however, the near infrared spectral range should be included. This wavelength range is particularly effective in distinguishing coniferous and broad-leaved forest and also in the biomass estimation [1, 3, 8].

In the future we will improve the radiometric corrections, make the mosaicking and DTM generation fully operative, and improve the methods of the actual numerical image interpretation.

References

1. T. Häme, "Landsat-Aided forest site type mapping", *Photogrammetric Engineering and Remote Sensing*, vol. 50, no. 8, pp. 1175–1183, 1984.
2. T. Häme, I. Heiler and J. San Miguel-Ayanz, "An unsupervised change detection and recognition system for forestry", *International Journal of Remote Sensing*, vol. 19, no. 6, pp. 1079–1099, 1998.
3. T. Häme, A. Salli, K. Andersson and A. Lohi, "A new methodology for the estimation of biomass of conifer-dominated boreal forest using NOAA AVHRR data", *International Journal of Remote Sensing*, vol. 18, no. 15, pp. 3211–3243, 1997.
4. R. M. Haralick, "Statistical and structural approaches to texture", in *Proceedings of the IEEE*, vol. 67, no. 5, pp. 786–804, 1979.
5. G. J. Hay, K. O. Niemann and G. F. McLean, "An Object-Specific image-texture analysis of h-resolution forest imagery", *Remote Sensing of Environment*, vol. 55, no. 2, pp. 108–122, 1996.
6. M. Holm and S. Rautakorpi, "First results of parallel global object reconstruction using digitized aerial photographs". *SPIE AeroSense'97 Conference on Integrating Photogrammetric Techniques with Scene Analysis and Machine Vision III*, Orlando, Florida, USA, 21–25 April 1997, pp. 165–175, 1997.
7. M. Holm, "New technology for making huge image mosaic maps and digital elevation models using parallel computing", *CSC News*, vol. 10, no. 1, pp. 21–23, 1998.

Forestry Applications of High Resolution Imagery

Fig. 3.1. A mosaic consisting of 141 images, and the corresponding DTM (2 × 7 km^2, 0.5 m resolution).

Fig. 3.2. Clustering result for the filtered, standard deviation images.

8. R. Kalensky and D. A. Wilson, "Spectral signatures of forest trees", in *Proceedings of the Third Canadian Symposium on Remote Sensing*, Edmonton, Alberta, September 22–24, pg. 18, Reprint, 1975.
9. , P. M. Narendra and M. Goldberg, "Image segmentation with directed trees", *IEEE Transactions on Pattern Analysis and Machine Intelligence*, vol. PAMI-2, no. 2, pp. 185–191, 1980.
10. E. Parmes, "Segmentation of Spot and Landsat satellite imagery", *Photogrammetric Journal of Finland*, vol. 13, no. 1, pp. 52–58, 1992.

Image Analysis Techniques for Urban Land Use Classification. The Use of Kernel Based Approaches to Process Very High Resolution Satellite Imagery

Charalambos C. Kontoes

Institute of Ionospheric and Space Research
National Observatory of Athens, Lofos Nimfon, 11810, Thission, Athens, Greece.
e-mail: kontoes@creator.space.noa.gr

Summary. Information on land use classes of urban areas is very important for their management and planning. Satellite remote sensing data can provide regular and up to date information on urban areas. Although satellite technology has changed significantly, a robust methodology in exploiting fruitfully satellite imagery in urban areas is under investigation.

This paper gives a literature review of image analysis techniques used for the classification of medium and high resolution satellite imagery in urban areas. In addition three kernel based classification techniques to process very high resolution imagery, are considered that integrate texture and spatial context properties for urban land use classes. The Evidence Based Interpretation Algorithm, a back-propagation neural network algorithm and the Kernel Reclassification algorithm. A summary of their use on Indian Remote Sensing (IRS) satellite imagery over the city of Athens is presented. Although some of these techniques return rather satisfactory results, the need for the development of more advanced interpretation methods, which integrate human reasoning in object identification is considered as indispensable, as the spatial resolution of the data is continuously increasing attaining the one of high altitude aerial photography (e.g. IKONOS system).

1. Introduction

The level of information that can be extracted from remotely sensed data is related to the spatial and temporal resolution of the acquired images. There is no doubt that the interpretation of the aerial photography typically provides a lot of the needed information for urban studies. However, this approach is costly and difficult to apply with sufficient frequency. On the other hand, satellite remote sensing can provide a method for acquiring regular and up to date information about urban areas, which may be particularly useful for monitoring changes within and on the fringes of urban development. Foster [7] makes an exhaustive presentation of the potentiality and advantages in using satellite technology for urban mapping and monitoring, but at the same time gives a detailed overview of the problems which arise when remote sensing imagery is treated in the frame of such studies.

During the last few years satellite technology has changed significantly and new sensors with higher spatial resolution and stereo-pair capabilities are offered for wide use (e.g. IRS-1C, IRS-1D, Radarsat, IKONOS, SPOT 5, KVR, KFA). But the technological advances have not solved the problem to identify a robust, transferable and easily repeatable methodology in exploiting fruitfully the satellite imagery in urban areas and many research topics are still open and under investigation. The first experiments in integrating satellite imagery in urban studies were not encouraging in terms of land use class specificity ([3, 30, 32]). This was initially attributed to the coarse spatial resolution of the satellite sensors. The fact that the spectral responses of different land cover types (e.g. buildings, roads, trees, grass, etc), which coexist in the field of view of the sensor, are averaged and registered as one pixel value, led to produce broad composite signals making thus difficult to distinguish between different land categories. Examples in using Landsat Multispectral Scanner (MSS) and Thematic Mapper (TM) data for urban classification are reported in various studies ([12, 19]). The only possible distinction achieved was between urban and non-urban areas. Other studies ([16, 23]) investigate the use of multi-spectral SPOT imagery or merged Landsat TM and SPOT Panchromatic data for urban classification. These studies also showed that spectral and spatial resolution of the SPOT data, were also inadequate for defining land cover/land use classes with the required specificity. The use of merged data returned slightly improved classification results but still far from the required detail of specificity.

The use of higher resolution data from recently developed sensors, although still under investigation, has not returned the expected improvements and in some cases the returned accuracy is even worse compared to the one achieved by the use of coarser resolution satellite data. This was attributed, to the problem of "scene noise" ([12, 22]), as the spectral signatures of urban areas are much more varied, since are composed by the spectral responses of individual scene elements. This makes land use class description and identification problematic, especially when classification is treated on a per-pixel fashion. In some cases, as in [6], the problem of "scene noise" is considered as being so unresolvable, that it is suggested to remove textural information and interclass variability from the scene plane, by the use of specific smoothing algorithms, in order to achieve better classification results. An overview of the experience in exploiting satellite imagery for urban land use mapping follows.

2. Satellite Image Interpretation Techniques for Urban Classification

The continuous advancement of satellite technology resulting in the acquisition of higher quality images, is strongly required for urban monitoring and change detection studies. As Barsnley and Barr [1] state, the problem to identify specific land use classes in urban areas, should not be attributed only to the spatial resolution of the data but mainly to the methods used to process and extract information from the scene. In many of the known studies the authors make use of pixel based approaches in classifying images of urban areas by the application of statistical cluster approaches. Although, this is a rather good technique for classifying large homogeneous agricultural fields, it may not return significant results within urban areas, especially because the spectral composition of urban land use classes, is violating the basic assumption for normal distribution ([11, 21]). This is because certain urban land use classes are composed by more land cover classes and with different percentage of their appearance within the boundaries of the land use class. Therefore, land use clusters for urban areas, will have multi-modal and considerably overlapping signatures which lead to significant miss-classifications. Therefore, pixel based approaches in urban areas, can suffer from being either too specific or too general, but far from the more abstract definition of land use classes.

The fundamental problem in producing accurate land use maps is that urban areas present a complex spatial arrangement of land cover types. Thus, the necessity for integrating techniques which account for the spatial arrangement of pixel values or alternatively land cover labels within a neighborhood was considered indispensable by a number of researchers. In many studies the spatial arrangement of pixel values is introduced by the use of texture measurements, which are integrated either in statistical classification models as additional layers or as input into advanced classification algorithms, through knowledge based systems, artificial neural network algorithms, image understanding approaches (especially the structural elements of texture). Texture has been used in the frame of many studies in the past, either in its structural or statistical form. Several investigators have used the most common gray-level co-occurence matrices to measure entropy, angular second moment, variance, correlation, etc. ([5, 8, 13, 14, 17, 27, 33]). Ryherd and Woodcock [29] introduce the "local variance" texture measure, which has been computed within a 3×3 adaptively placed window is introduced for image segmentation. The method was tested on satellite data representing rural and urban areas. The experiments showed that the combined use of spectral with texture information returned in all the cases improved classification accuracy especially within urban areas exhibiting pronounced texture characteristics. Paola and Schowengerdt [26] have devised an interesting way to provide spatial texture to the classification approach. In this study a 3×3 geometric window of pixels is considered as input to a neural net work classification approach. The method was tested on a Landsat TM scene representing a complex urban area. The results were satisfactory since the incorporation of texture resulted in the reduction of the required time for network training and at the same time the returned thematic accuracy, was higher than the one given also by a neural net but based on a pixel-by-pixel approach instead of a 3×3 window.

124   C. C. Kontoes

Except for the statistical measurements of texture the structural methods have been also used in the frame of relevant image interpretation and analysis studies. Nagao and Matsuyama [25] introduce image understanding systems which is divided into two sub-systems: the low level which measures and assigns specific geometric properties to objects, which have been given by image segmentation. In the high level processing which follows, the use of specific knowledge, through an expert system, results in object recognition on the basis of their geometric properties and spatial relations with other neighboring objects. This detection system, was developed to recognise objects like, two types of crop fields, bare soil, crop field without plant, forest, grassland, road, river, car, building, house. In a very similar way Gahegan and Flack [9], have experimented with image understanding techniques integrating structural, spectral and spatial properties to identify, through the use of a knowledge based system, the labels of specific objects. Their system encompasses both pixel based but also feature based processing techniques to extract significant information for image interpretation.

However, the classification in urban land use classes is more than the identification of the exact label of classified objects on the image plane. Urban land use classes present a complex spatial assemblage of land cover types introduced by pixel labels or entire region labels, which are usually the result of the image classification and/or segmentation. A technique which accounts both for the frequency and spatial arrangement of class labels within a square kernel has been applied by Barsnley and Barr [1]. This technique was first suggested by Wharton [34] and divides the classification process into two stages: during the first stage each pixel is assigned a certain classification label through a standard per pixel classification approach. The second applies a spatial post classification process to measure the frequency and spatial arrangement of pixel labels within a user defined kernel to infer urban land use classes. The technique is using a statistical similarity criterion to assign the central pixel of the kernel to a land use class. In the study of Barsnley and Barr [1] initial classifications of urban areas in land cover types such as "small structure", "large structure", "tree", "crop", "grass", and "soil" were combined spatially through the use of the kernel based reclassification technique, in order to infer land use classes like "low-density residential", "medium density residential", "commercial", etc.

The integration of ancillary data in an attempt to refine and improve the classification in urban areas, is realised either during classification or in a post-classification stage of the process. There are two approaches to using the ancillary data during classification. The first introduces the ancillary layer as an additional layer forming an extended vector of values for each pixel ([31]). The second makes use of the ancillary layer to modify the apriori probabilities of the classes under consideration before the employment of the classification algorithm. The latter technique has been proved very demanding in terms of sampling efforts and the quality of the classification output is very scene depended. Moreover, the integration of the ancillary data as component of an extended vector usually results in the violation of the normal distribution of pixel values, required by many statistical classification approaches. These remarks have led many researchers to make use of ancillary data into a post-classification process. Kontoes *et al.* [20] has integrated ancillary thematic layers introducing geographic and spatial context information, through a knowledge based system to refine the first classification output. Another typical example is the work of Harris and Ventura [15]. In this study the integration of zoning and housing density data and their use in the frame of a knowledge based system, resulted in the reduction of confusion between classes and the increase of the number of identifiable land use classes within an urban area. The initial classification in 5 land use classes was further refined into 15 classes by integrating the two additional layers. Similarly Moller-Jensen [24] is integrating texture and context information into a knowledge-based classification approach to classify urban areas.

3. Kernel Based Classification Approaches for the Interpretation of IRS-1C Satellite Imagery

On the basis of the above mentioned studies, it becomes clear that the objective to define adequate image analysis techniques, using statistical and non-statistical approaches to extract reliable signatures for urban land use classes, requires the development of algorithms which measure the spatial distribution of the spectral and texture properties found on the pixel context and produce significant features to introduce into the classification procedure. Currently the author is involved into a pilot project, funded by the European Commission, aiming to exploit very high resolution imagery provided by the Indian Remote Sensing, IRS-1C, satellite system for urban classification of the city of Athens, according to the Eurostat's CLUSTERS[1] nomenclature scheme. Three Kernel Based Classification Approaches have been considered, for the purposes of the study. The EBIS (Evidence Based Interpretation of Satellite Images), the supervised and unsupervised artificial neural network (ANN) and the kernel reclassification algorithms.

The application of an unsupervised neural network based on a topological network (Kohonen map) is suggested in order to decide about the optimal kernel size and input image information content, similarly to the work of Paola and Schowengerdt [26] and Wilkinson *et al.* [35]. The output of this investigation, is used to feed appropriately the back-propagation neural network, consisted of one input layer, two hidden layer(s) and an output layer. The number of hidden layers and their dimension is chosen heuristically, based on previous experience and studies.

---

[1] Classification for Land Use Statistics: Eurostat Remote Sensing programme.

The "Evidence-Based Interpretation of Satellite Images" algorithm (EBIS) makes use of the reasoning scheme of the Shafer's theory of evidence ([4, 10]). Therefore evidence functions are used to decide whether a pixel is assigned to a certain class or not. The parametric model used by the algorithm, is the multinomial distribution. Via evidence functions the pixel is assigned to a special class in case that the hypothesis, which is the local histogram of the training area, matches the histogram of the pixel observed in a specific window. The evidence function measures the degree to which a pixel's feature match a class hypothesis. If the pixel's feature match a histogram of a training area well, then this constitutes evidence in favour of the corresponding class hypothesis. Apart from the local histogram, EBIS also supports the common co-occurrence matrices introduced by Haralick *et al.* [13]. This is because, there are classes, which show the same histogram, but are clearly distinguishable by the human eye (e.g. an image with black and white stripes has the same histogram as an image with the same number of black and white pixels arranged as the pattern on a chess-board). In this case co-occurrence matrices, which record the relative frequencies of spatial co-occurrences of grey values, are useful. Horizontal, vertical, left-diagonal and right-diagonal co-occurrences can be defined.

The Kernel Based Reclassification Algorithm attempts to derive information on urban land use based on the frequency and the spatial arrangement of the land cover labels within a square kernel. The assumption underlying this approach is that individual categories of land use have characteristic spatial mixtures of spectrally distinct land cover types that enable their recognition in high spatial resolution images ([2, 34]).

The employment of neural networks ([18, 36]), evidential based approaches ([4, 10, 28, 33, 36]) and reclassification techniques, allows the integration of any type of sources of data classes to derive information (land use) classes, independently if the assumption of normal distribution of density functions is fulfilled or not. At the same time these techniques introduce texture and contextual information into the classification procedure.

4. Evaluation of the Use of the Kernel Based Algorithms to Classify IRS-1C Satellite Data over Athens

The application of the three kernel based classification algorithms on high resolution IRS-1C satellite data over Athens, provided useful information concerning the classification abilities of texture based classifiers in respect to the Eurostat's CLUSTERS nomenclature scheme used for land use description. It should be noted that the CLUSTERS nomenclature scheme presents four hierarchical levels of detail in land use/land cover classes for urban, agricultural and forested areas. The experiments showed that some urban land use classes at Level VI and Level III could be derived with relatively good accuracy by applying kernel based classifiers. However, this was the case only for the neural network (figure 4.2) and kernel reclassification algorithms (figure 4.3). The use of EBIS classifier did not produce meaningful map products. The EBIS algorithm returned the best results when the only input layer was the panchromatic scene of IRS-1C. In contrast multi-temporal and multi-spectral satellite data resulted in noisy classification products. The neural network and kernel reclassification algorithms produced more accurate maps at the fourth Level of the CLUSTERS nomenclature scheme and four residential classes which differ from one another in terms of housing density, were returned. These classes are "Continuous dense residential", "Continuous residential with moderate density", "Discontinuous residential with moderate density", "Isolated building areas". However these classes, are only a subset of the CLUSTERS classes, as many of the other classes in the nomenclature are functional classes, which can be derived solely from ancillary data and analog photo-interpretation. Although many experiments have been conducted by using various kernel sizes and image layer combinations, including multi-spectral and multi-temporal sets of enhanced (fused) data, it was not possible to separate class "Industrial or commercial activities". This most likely happens because there does not exist a typical texture pattern for this class on the image plane.

The best classification results showed an accuracy of the order of 72–75% for the various classified urban classes. In general, the texture based classification approaches proved to be useful for the classification of high resolution data. In contrast, pixel-based approaches like e.g. the maximum likelihood classifier, classify most of these classes as a mixture of single classes, like vegetation, roads and houses of different kinds resulting in confused classifications with a lot of "salt and pepper" noise (figure 4.1). The results showed, that the classification accuracy is directly linked to the applied size of the kernel used, since heterogeneous classes need large window sizes to be correctly represented, whereas small objects demand the contrary. Therefore, a combination of both results is necessary in case homogeneous classes are needed, but without accepting a loss of all small structures.

Further research is required to study a set of pre-defined kernel sizes for data classification, which depend on the class or set of classes to be classified. For the purposes of this study, a trial and error approach was followed in order to decide about the best window size to be applied for the IRS-1C data classification. In all the experiments, the texture properties of the land use classes have been introduced by the use of enhanced (merged) satellite data, which is the result of the integration of multi-spectral images of the LISS III sensor with the high spatial resolution panchromatic image of the IRS-1C Panchromatic sensor.

The following sections describe in greater detail the classification results obtained from the maximum likelihood, neural network and kernel reclassification approaches.

4.1 Maximum Likelihood Classification

Five classification experiments have been conducted within the urban area of the city of Athens. Various combination of input layers, comprising multi-temporal and enhanced multi-spectral IRS-1C LISS III data with the addition of texture (variance layer) have been considered as input to the classification procedure. The aim was to evaluate the degree by which a per-pixel classification approach based on spectral and textural properties could return meaningful classifications. The experiments showed that only non-urban classes could be classified accurately. With the only exception of class "Dense Residential Areas" (A111) the rest of the urban land use classes (A) were classified wrongly and a lot of inter class confusion revealed. The classification accuracy for urban classes was ranging between 20 to 45%. Only at mixed levels of the nomenclature scheme the returned accuracy values are at higher levels. Figure 4.1 illustrates a subset of 514 lines × 319 pixels out of the total study area ($40 \times 40\ km^2$).

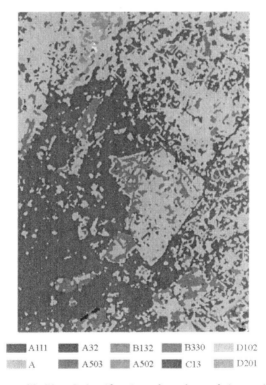

Fig. 4.1. Maximum likelihood classification of a subset of the study area. Classification nomenclature: A = 'man-made areas', A111 = 'continuous dense residential', A32 = 'transport', A502 = 'sport facilities', A503 = 'green or leisure areas', B132 = 'fresh vegetation', B3 (B330) = 'permanent crops', C13 = 'conifers', D102 = 'bushes', D201 = 'herbaceous vegetation'.

4.2 Neural Network Classification

The back-propagation neural network classifier was applied on the combination of multi-temporal very high resolution and enhanced IRS-1C LISS III data. The neural network consisted of four layers: one input, two hidden and one output layer. The network applies image classification within a 7 × 7 kernel scanning the image. The experiment showed that the application of the back-propagation neural network resulted in accurate discrimination between urban land use classes as a function of housing density within the urban area of the city of Athens. The overall classification accuracy achieved was 72,02%. Figure 4.2 illustrates a subset of 514 lines × 319 pixels out of the total study area (40 × 40 km^2).

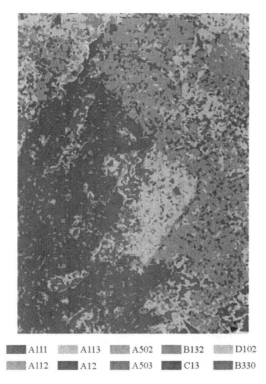

Fig. 4.2. Neural network classification of the same subset of the study area as in figure 4.1. Classification nomenclature: As in figure 4.1 with the addition of, A112 = 'continuous residential of moderate density' A113 = 'discontinuous residential of moderate density', A12 = 'public services'.

4.3 Kernel Reclassification Algorithm

The kernel reclassification was applied on the first classification map, resulted by the employment of the ISODATA clustering algorithm on the enhanced IRS-1C LISS III satellite data. The unsupervised classification layer was introducing five land cover classes. The kernel reclassification algorithm inferred land use classes on the basis of the spatial arrangement of the five land cover labels within a 11×11 kernel. Similarly to the neural network, the experiment showed that this approach may result in accurate discrimination between urban land use classes as a function of housing density within the urban area of the city of Athens. The overall classification accuracy achieved was 72,04%. Figure 4.3 illustrates a subset of 514 lines × 319 pixels out of the total study area ($40 \times 40\ km^2$).

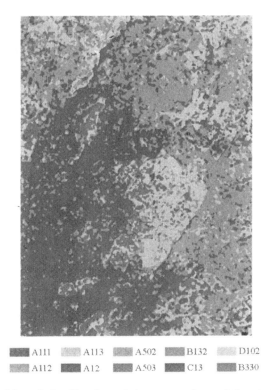

Fig. 4.3. Kernel based classification of the same subset of the study area. Classification nomenclature as in figure 4.2.

5. Conclusions

From the three classification experiments it was concluded that the back-propagation neural network classification returned a rather good classification, in terms of class specificity and accuracy. The kernel reclassification algorithm could be considered as producing the same level of accuracy for the most of the urban classes. The kernel reclassification algorithm presents the same potentiality for residential class discrimination as the neural network approach. The only exception is class "Industrial and/or commercial". In general, this class was given with low classification accuracy, since there is no a typical texture pattern representing industrial and commercial zones. The rest of the classes are characterised either by their function (e.g., cultural sites, technical network infrastructure, manufacturing industry, heavy industry, marine transports, etc) or structure (e.g., airport, sport facilities, etc.). These land use classes may not be classified by automatic procedures integrating texture and spectral properties. There is a need of ancillary information and visual photo-interpretation or computer vision approaches to be applied. The analysis of image structure, which has been studied on aerial photographs by other scientists, integrating geometric properties, empirical and heuristic knowledge, semantic knowledge and computer vision techniques, should be considered thoroughly, for urban land use classification, especially when very high spatial resolution satellite imagery is used.

References

1. M. J. Barnsley and S. L. Barr, "Inferring urban land use from satellite sensor using kernel-based spatial reclassification", *Photogrammetric Engineering and Remote Sensing*, vol. 62, no. 8, pp. 949–958, 1996.
2. M. J. Barnsley and S. L. Barr, "Developing kernel-based spatial reclassification techniques for improved land use monitoring, using high spatial resolution images", in: *Proceedings XXIX Conference of the International Society for Photogrammetry and Remote Sensing (ISPRS'92), International Archives of Photogrammetry and Remote Sensing: Commission 7*, 2–14 August 1992, Washington DC, pp. 646–654, 1992.
3. M. J. Barsnley, G. J. Sadler and J. S. Shepherd, "Integrating remotely sensed images and digital map data in the context of urban planning", in: *Proceedings of the 15th Annual Conference of the Remote Sensing Society*, Bristol, UK, pp. 25–32, 1989.
4. J. A. Barnett, "Computational methods for a mathematical theory of Evidence", *Proceedings of the Seventh International Joint Conference on Artificial Intelligence*, Vancouver, BC, pp. 868–875, 1981.
5. R. W. Conners and C. A. Harlow, "A theoretical comparison of texture algorithms", *IEEE Transactions on Pattern Analysis and Machine Intelligence*, vol. PAMI-2, no. 3, 1980.
6. J. L. Cushnie, "The interactive effect of spatial resolution and degree of internal variability within landcover types on classification accuracies, *International Journal of Remote Sensing*, vol. 8, pp. 15–29, 1987.

132   C. C. Kontoes

7. B. C. Foster, "An examination of some problems and solutions in monitoring urban areas from satellite platforms", *International Journal of Remote Sensing*, vol. 6, pp. 139–151, 1985.
8. M. M. Galloway, "Texture analysis using grey level run lengths", *Computer Graphics and Image Processing*, vol. 4, pp. 172–179, 1992.
9. M. Gahegan and J. Flack, "A model to support the integration of image understanding techniques within a GIS", *Photogrammetric Engineering and Remote Sensing*, vol. 62, no. 5, pp. 483–490, 1996.
10. J. Gordon and E. H. Shortliffe, "A method for managing evidential reasoning in a hierarchical hypothesis space", *Artificial Intelligence*, no. 26, pp. 323–357, 1985.
11. E. S. Gilbert, "The effect of unequal variance-covariance matrices on Fisher's linear discriminant functions", *Biometrics*, vol. 25, pp. 505–516, 1969.
12. B. Haac, N. Bryant and S. Adams, "An assessment of Landsat MSS and TM Data for urban and near urban land cover digital classification", *Remote Sensing of Environment*, vol. 21, pp. 201–213, 1987.
13. R. M. Haralick, K. Shanmugam and I. Dinstein, "Textural features for image classification", *IEEE Transactions on Systems, Man, and Cybernetics*, vol. SMC-3, no. 6, pp. 610–621,1973.
14. R. M. Haralick, "Statistical and structural approaches to texture", *IEEE Proceedings*, vol. 67, pp. 786–804, 1979.
15. P. M. Harris and S. J. Ventura, "The integration of geographic data with remotely sensed imagery to improve classification in an urban area", *Photogrammetric Engineering Remote Sensing*, vol. 61, no. 8, pp. 993–998, 1995.
16. A. R. Harrison and T. R. Richards, "Multispectral classification of urban land use using SPOT HRV data", *Digest International Geoscience and Remote Sensing Symposium*, Edinburgh, UK, pp. 205–206, 1988.
17. J. Jensen, "Spectral and textural features to classify exclusive land cover at the urban fringe", *The Professional Geographer*, vol. 4, pp. 400–409, 1979.
18. I. Kanellopoulos, G. G. Wilkinson and J. Mégier, "Integration of neural network and statistical image classification for land cover mapping", in: *Proceedings of International Geoscience and Remote Sensing Symposium, IGARSS'93*, Tokyo, pp. 511–513, 1993.
19. S. Khorram, J. A. Brochaus and H. M. Chesire, "Comparison of Landsat MSS and TM Data for urban land use classification", *IEEE Transactions on Geoscience and Remote Sensing*, vol. GE-25, no. 2, pp. 238–243, 1987.
20. C. C. Kontoes, D. Rokos, G. G. Wilkinson and J. Mégier, "The use of expert system and supervised relaxation techniques to improve SPOT image classification using spatial context", in: *Proceedings International Geoscience and Remote Sensing Symposium, IGARSS '91*, vol. 3, pp. 1855–1858, 1991.
21. S. Marks and J. O. Dunn, "Discriminant functions when covariance matrices are unequal", *Photogrammetric Engineering and Remote Sensing*, vol. 69, pp. 555–557, 1974.
22. L. R. G. Martin, P. G. Howarth and G. Holder, "Multispectral classification of land use at the rural-urban fringe using SPOT data", *Canadian Journal of Remote Sensing*, vol. 14, pp. 72–79, 1988.
23. L. R. G. Martin and P. G. Howarth, "Change detection accuracy assessment using SPOT multispectral imagery of the rural-urban fringe", *Remote Sensing of Environment*, vol. 30, pp. 55–66, 1989.
24. L. Moller-Jensen, "Knowledge-based classification of an urban area using texture and context information in Landsat TM imagery", *Photogrammetric Engineering and Remote Sensing*, vol. 56, no. 6, pp. 899–904, 1990.

25. M. Nagao and T. Matsuyama, *A structural analysis of complex aerial photographs*, Plenum Press, New York, pp. 190, 1980.
26. D. J. Paola and R. A. Schowengerdt, "The effect of neural network structure on a multispectral land use/land cover classification", *Photogrammetric Engineering and Remote Sensing*, vol. 63, no. 5, pp. 535–544, 1997.
27. D. R. Peddle and S. E. Franklin, "Image texture processing and data integration for surface pattern discrimination", *Photogrammetric Engineering and Remote Sensing*, vol. 57, no. 4, pp. 413–420, 1991.
28. D. R. Peddle, "Knowledge formulation for supervised evidential classification", *Photogrammetric Engineering and Remote Sensing*, vol. 61, no. 4, pp. 409–417, 1995.
29. S. Ryherd and C. Woodcock, "Combining spectral and texture data in the segmentation of remotely sensed images", *Photogrammetric Engineering and Remote Sensing*, vol. 62, no.2, pp. 181–194, 1996.
30. G. J. Sadler and M. J. Barsnley, "Use of population density data to improve classification accuracies in remotely sensed images of urban areas", in: *Proceedings of the 1st European Conference on GIS (EGIS 90)*, Amsterdam, EGIS foundation, Utrecht, pp. 968–977, 1990.
31. J. D. Spooner, "Automated urban change detection using scanned cartographic and satellite Image data", Technical Papers, ACSM-ASPRS Fall Convention, Atlanta, Georgia, pp. B118–B126, 1991.
32. D. L. Toll, "Effect of Landsat Thematic Mapper sensor parameters on land cover classification", *Remote Sensing of Environment*, vol. 17, pp. 129–140, 1985.
33. F. M. Vilnrotter, R. Nevatia and K. E. Price, "Structural Analysis of Natural Textures", *IEEE Transactions on Pattern Analysis and Machine Intelligence*, vol. PAMI-8, 1986.
34. S. W. Wharton, "A contextual classification method for recognizing land use patterns in high resolution remotely-sensed data", *Pattern Recognition*, vol. 15, pp. 317–324, 1982.
35. G. G. Wilkinson, C. C. Kontoes and C. N. Murray, "Recognition and inventory of oceanic clouds from satellite data using an artificial neural network technique", *International Symposium on Dimethylsulphide Oceans Atmosphere and Climate*, DG XII/E CCE, Belgirate, Italy, 1992.
36. G. G. Wilkinson, I. Kanellopoulos, C. C. Kontoes and J. Mégier, " A comparison of neural network and expert system methods for analysis of remotely-sensed imagery", in: *Proceedings of International Geoscience and Remote Sensing Symposium, IGARSS'92*, Houston, vol. I, pp. 62–64, 1992.

Part III

Visualisation, 3D and Stereo

Automated Change Detection in Remotely Sensed Imagery

Joseph Mundy and Rupert Curwen

G. E. Corporate Research and Development
1 Research Circle, Niskayuna, NY 12309, USA

Summary. A major application of remotely sensed imagery is the detection of changes in the configuration of man-made and natural structures. Classical multispectral imagery analysis has primarily relied on pixel-level classification to detect such changes. In this presentation, a system called FOCUS will be described which detects changes based on spatial analysis in the context of a 3-dimensional model of the world. The algorithms employed by FOCUS assume the availability of extensive context about sensor photogrammetric parameters and illumination. With such context, relatively simple image segmentation and model matching procedures can yield effective analysis of change.

1. Introduction

Before considering technical solutions for change detection it is essential to describe more fully the problem to be solved. The key issue is the definition of *change*. This paper is primarily concerned with changes of military importance although there is considerable overlap with the observation of changes of commercial interest.

Military intelligence is the process of gathering information to permit an accurate assessment of an enemy's ability to wage war and to determine its plans for aggression and current state of readiness. This information is vital for strategic planning and to provide essential data to direct tactical actions. Once the intelligence-gathering process is considered from this broad perspective, it is clear that change, is as diverse as all of human economic activity.

For example, consider the process of weapon deployment, such as a field artillery piece. The employment of this weapon in an act of aggression is a final event which should be avoided or prevented if possible. The role of intelligence is to investigate the entire chain of events leading up to the final, military deployment and use of the weapon. This event sequence is typically long and complex, ranging from mining of the raw materials, testing and design evolution, to transport and supply. This range of processes is typical of economic activity in general and represents a vast network of human and material interaction.

Only a small portion of such events can be directly observed from overhead imagery. The intelligence process builds a picture of the overall state of affairs from discrete observations such as:

138   J. Mundy and R. Curwen

- Presence or absence of road vehicles/transporters
- Presence or absence of railcars
- Presence or absence of aircraft
- Occupancy of material storage areas, particularly;
  - Oil drums
  - Chemical drums
  - Weapon shipment containers
- Construction
  - Earth movement
  - Concrete forms
  - Scaffolding

These visible changes can be used as clues in formulating an overall assessment using knowledge of the underlying processes. In the remainder of the paper, algorithms designed to detect such changes will be described.

2. Context-Based Algorithms

A key premise in the approach taken here is that typical computer vision algorithms, such as texture classification or model-based recognition are inherently too error prone to produce reliable change detection. It is not that the algorithms fail to perform according to the assumptions upon which the algorithm design is based. Instead, it is the assumptions themselves which are at fault.

For example, in matching a 3-dimensional polyhedral model of an object in an image it is assumed that the projected visible edges of the object in the image will be largely recovered by image segmentation. The segmentation process typically consists of edge detection followed by line fitting. In the case of poor image contrast or occlusion the desired object boundaries will not be present. An example of this difficulty is illustrated in figure 2.1, which is an image of two aircraft taken under foggy conditions. In figure 2.2, two of the aircraft are segmented using a modified Canny edge detector [1]. The parameters of the edge detection were adjusted for best feature recovery. It is clear that much of the potential projected aircraft boundary is not adequately detected in either figure 2.2 a) or b). Suppose however, that one could use extra information derived from the appearance of other fixed structures in the same scene under the same conditions to compensate for such inadequate segmentation results.

One example of such fixed structures are the painted marks on the dark tarmac surface situated above the aircraft parking areas. The segmentation of these marks, along with an aircraft is shown for two conditions in figure 2.3. Similarly to the aircraft, few of the tarmac marking edges are detected in the cloudy image compared with the clear, sunny conditions. The ratio of

Automated Change Detection in Remotely Sensed Imagery   139

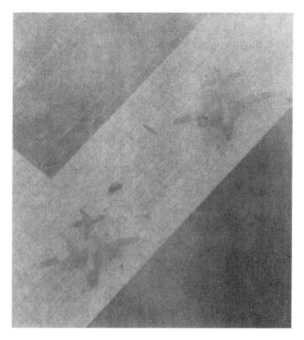

Fig. 2.1. Two aircraft with a gradient of fog overlaying the scene.

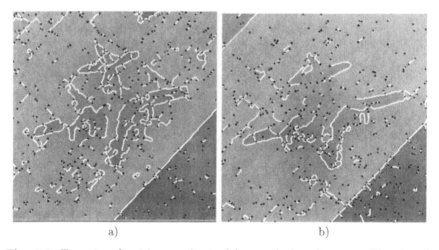

Fig. 2.2. Two aircraft with a gradient of fog overlaying the scene. The aircraft shown are adjacent on a parking apron. Detected boundaries are shown as white curves with boundary junctions as black dots. The first aircraft in a) produces a highly fragmented boundary. The visible boundary of the second aircraft in b) is even more sparse.

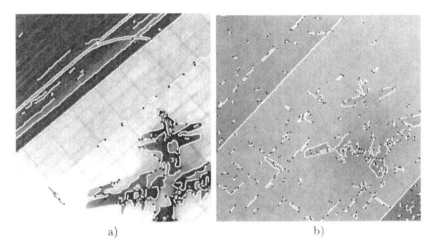

Fig. 2.3. a) The segmentation of the runway area on a clear day using similar edge detection parameters to figure 2.2. b) The same scene under cloudy conditions.

recovered edge perimeter can be used to modulate the expected matched boundary.

These fixed reference features can be encoded in a 3-dimensional site model[1]. Such a model is illustrated in figure 2.4 a). The model is projected into the image using the known sensor projection transformation. The registration of such a projection can be refined by matching the known structures to their corresponding segmented image features.

The edge boundary calibration approach just cited is an example of *context* derived from the site model and known sensor characteristics to condition the algorithm behavior. This approach is called context-based vision, or sometimes known as model-supported exploitation. The approach was first suggested by Strat and Fischler [5]. In their system, called Condor, context about the 3-dimensional world was used to condition hypotheses during the interpretation of 2-dimensional scenes.

This concept was extended to provide the basis for a five year project by the US Defence Advanced Projects Research Agency (DARPA). The project, called RADIUS, centered on the use of 3-dimensional site models to guide the application of algorithms for change detection. A considerable number of universities and other research institutions were involved and an extensive testbed was constructed to evaluate the concept of context. A set of final overview reports has been recently published [2].

At the end of the RADIUS project, General Electric CRD in partnership with Lockheed-Martin Missiles and Space extended the RADIUS technology

---

[1] See the chapter by Moons *et al.*, page 148, in this book, which describes methods for constructing such models from multiple image views.

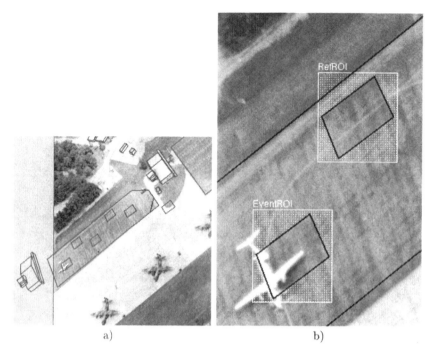

Fig. 2.4. a) A 3-dimensional site model projected into an image. Note that there are small polygonal regions placed over the center of aircraft parking regions. b) A close-up view of the projected edge detection regions of interest which are defined by the site model.

to produce a system called FOCUS. This system and some of the algorithms employed within the FOCUS system will now be described.

3. Context-Based Change Detection

One can apply context at various levels of image representation. In this section, examples illustrating change detection for two representational levels are presented.

3.1 Change Based on Step Edge Detection

A very effective image segmentation process is to detect all positions in the image which match an intensity model for step image discontinuities. A standard step edge detection algorithm is the Canny [1] algorithm. An example of this segmentation was shown in figure 2.2.

This image representation can then be used to detect changes at specific spatial locations, by extracting step edges for the same 3-dimensional surfaces

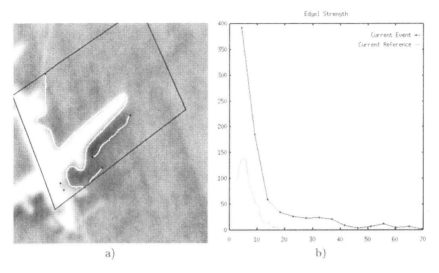

Fig. 3.1. a) The image step edge segmentation within the region of interest. b) Histograms of step edge gradient strength for the event and reference regions.

across a series of images. The segmentation is carried out within projected 3-dimensional polygonal regions as shown in figure 2.4. Suppose the goal is to detect the presence or absence of aircraft in the fixed parking areas within the darker tarmac area between the two hangar buildings. The problem is relatively easy because it is required for the aircraft to park at well-defined locations indicated by the parking marks on the tarmac. These locations are bounded by 3-dimensional polygons, constructed manually using multiple image views of the site.

Using polygon scan conversion, the edge detection algorithm can be applied strictly within the indicated regions of interest. In the figure, two regions are shown. The left region, defined over the small aircraft is called the event region. The right region, defined over an empty tarmac surface is called the reference region.

In the algorithm, only the gradient strength of the detected step edges is considered. The histograms of the gradient of image intensity, a measure of step strength are shown in figure 3.1. These histograms are the basis for change detection. The rationale for this choice is that objects of interest generally produce sharp contrast edges or shadows which produce high contrast boundaries. On the other hand, background material such as concrete or tarmac appears with much lower contrast.

This relationship is illustrated in figure 3.1 where the gradient strength is shown for the background reference region and the high contrast aircraft. Note that the peak gradient contrast for the aircraft is more than three times higher than the background tarmac.

Automated Change Detection in Remotely Sensed Imagery   143

For this algorithm, change is measured in terms of the characteristics of these two histograms. Work by Pietikainen *et al.* [4] has demonstrated that an effective distance measure between histograms should account for the entire distribution, rather than a parametric approximation of it. The distance metric is defined on two distribution functions, r_i a histogram of the reference region and e_i the histogram for an event region where change has occurred. The change measure is defined as:

$$D(r,e) = \frac{\sum_i^n \left[r_i \log(\frac{r_i}{e_i}) - e_i \log(\frac{e_i}{r_i}) \right]}{\sum_i^n [r_i + e_i]}.$$

This measure is slightly different from that used by Pietikainen but both are derived from a log likelihood class membership test. $D(r,e)$ measures the similarity of two histogram distributions, the larger the magnitude of $D(r,e)$, the more likely that the two histograms are drawn from a different population. $D(r,e)$ can be generalized for comparing joint distributions as well as distributions of a single parameter.

Because the change is precisely delineated by the event and reference regions, an algorithm with this simplicity can achieve reasonably robust results. Invariance to intensity and image quality provided by the reference region also is a major factor in the success of the approach.

3.2 Change Based on Geometry

As a second example, the use of 3-dimensional object models is considered. In this approach, a geometric model of an object is constructed, perhaps from multiple images or perhaps from drawings. An example is shown in figure 3.2. There are many algorithms to match a 3-dimensional model to 2-dimensional geometry derived from image segmentations. The key here is to constrain the matching process using the context of the overall site model. In the aerial image application, we usually know a great deal about the image sensor in terms of external and internal orientation.

Carrying forward with the detection of aircraft in fixed parking locations, the uncertainty in location and orientation is relatively small, but there can be high variation in the type of aircraft parked in a given location. If the change of interest involves the type of aircraft, rather than just simple presence/absence, then a more specific representation of the object, such as 3-dimensional geometry, much be employed.

The 3-dimensional model is projected in an image and matched against corresponding image boundary segments. The Hausdorff matching algorithm, developed by Dan Huttenlocher [3]. The algorithm uses a measure of mismatch between two point sets called the Hausdorff metric, which defines a worst case distance between two sets of points, S_1, and a set of points, S_2. The directed Hausdorff distance is defined as:

$$h(\mathcal{S}_1, \mathcal{S}_2) = Max_j Min_{j>k} d(\mathbf{p}_j, \mathbf{p}_k). \qquad (3.1)$$

where $d(\mathbf{p}_j, \mathbf{p}_k)$ is the standard Euclidean distance between two points, $\mathbf{p}_j$, $\mathbf{p}_k$.

The algorithm operates by adjusting the a translation vector between the two point sets which minimizes the Hausdorff error. In the current implementation, a number of solutions are obtained corresponding to the percentile of points satisfying a bound on Hausdorff distance. That is, a match is considered feasible if some specified percentage, f, of the points lie within a Hausdorff distance, $h_{\max}$. It has been determined that a distance of two pixels give satisfactory tolerance to normal variations in the edge smoothness of object boundaries. With adequate segmentation, a fraction, $f = 0.6$, of the projected boundary can be expected. An example run of the algorithm is shown in figure 3.3. In the example, there are three significant changes detected, which are shown near the top of the figure. However, one of the detected changes is incorrect. There is actually a C130 present, but the edge detection failed to find enough edges to match the model. The image conditions for case were shown earlier in figure 2.1. As described above, this error can be eliminated by taking into account the expected percentage of boundary match, using other fixed site features.

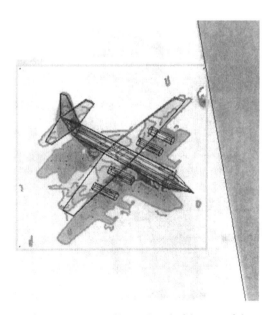

Fig. 3.2. An example of matching a 3-dimensional object model to a set of image segmentation boundaries. The model is shown as dark lines. The segmentation is rendered in a middle grey tone. In this case, the match between the model and the segmentation boundaries was achieved using a translation search based on the Hausdorff distance metric.

Fig. 3.3. An example run for the Hausdorff matching algorithm. The goal is to detect a missing C130 aircraft.

4. Discussion

The FOCUS change-detection system embodies six algorithms which are of similar philosophy to the two described above. Descriptions used in these other algorithms include: direct image intensity, orientation of line segments, distribution of edge boundary orientation, 3-dimensional boundary symmetry. In all cases, these algorithms are applied in a limited spatial context using the known sensor and illumination attributes.

The system has been implemented with a web-browser interface so that change detection can be carried out on a central server and change results accessed by many distributed users. The web interface also provides tools for creating and editing new change detectors. As imagery for a given site comes into the server, it is processed by applicable and active change detectors. The

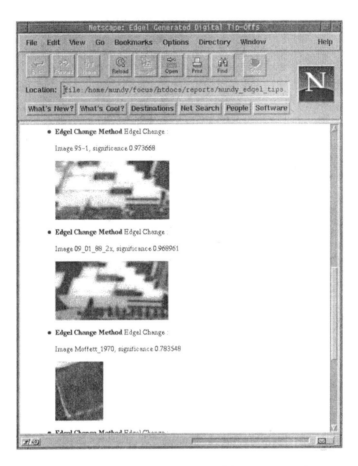

Fig. 4.1. A sample tip-sheet produced by the FOCUS system. Here as search has been carried out through historical imagery to find an image where a building does not yet exist. This process is called negation.

results are sorted into a web page, called the tip-off sheet. An example is shown in figure 4.1.

It may seem that these algorithms and the context in which they operate are overly contrived. However, as was demonstrated many times during the workshop, most remote sensing algorithms have achieved at best 80% success. Given this state of affairs, for automated analysis to become of practical interest, it will be necessary to define the structure of the application to take into account these limitations.

The use of very constrained context illustrated here is one way forward. It is still the case that even such constrained algorithms can fail. The use of a web-browser and search engine protocol ameliorates such failures since the user only expects a reasonable sorting of the changes, not a perfect assignment of change classification.

References

1. J. F. Canny, "Finding edges and lines in images", Technical Report AI-TR-720, Massachusetts Institute of Technology, Artificial Intelligence Laboratory, June 1983.
2. O. Firschein, ed., *RADIUS: Image understanding for imagery intelligence.* San Francisco: Morgan Kaufmann, 1997.
3. Daniel P. Huttenlocher, G. Klanderman, and W. Rucklidge, "Comparing images using the Hausdorff distance", *IEEE Transactions on Pattern Analysis and Machine Intelligence*, vol. 15, pp. 850–863, October 1993.
4. T. Ojala and M. Pietikainen, "Performance evaluation of texture measures with classification based on kullback discrimination of distributions", in *Proceedings, 12th International Joint Conference on Pattern Recognition*, Jerusalem, Israel, pp. 582–585, October 1994.
5. T. Strat and M. Fishler, "Context-based vision: Recognizing objects using information from both 2-d and 3-d imagery", *IEEE Transactions on Pattern Analysis and Machine Intelligence*, vol. 13, no. 10, pp. 1050–1066, 1991.

A 3-Dimensional Multi-View Based Strategy for Remotely Sensed Image Interpretation*

Theo Moons, David Frère, and Luc Van Gool

Katholieke Universiteit Leuven, ESAT – PSI,
Kard. Mercierlaan 94, B-3001 Heverlee, Belgium.
e-mail: {Theo.Moons,David.Frere,Luc.VanGool}@esat.kuleuven.ac.be

Summary. The aim of this paper is to assess the feasibility of extracting 3-dimensional information about man-made objects from very high resolution satellite imagery. To this end, a 3-dimensional, multi-view based paradigm is proposed. The underlying philosophy is to generate from the images reliable geometric 3D features, exploiting the multi-view geometric constraints for blunder correction. Scene analysis is performed by reasoning in 3D world space and verified in the images by means of a hypothesis generation and verification procedure in which the decisions are taken on the basis of a multi-view consensus. As an example of such an approach, a method for automatic modelling and 3D reconstruction of buildings is discussed and the effects of the resolution of the new generation satellite data on the performance of the algorithm and the metric accuracy of the final reconstruction is investigated.

1. Introduction

The new generation of satellite imaging sensors opens new perspectives for remote sensing applications. First of all, the *very high resolution* of the image data (viz. $80\,cm - 2\,m$) permits the use of *geometric* image *features* such as points, lines, curves, etc. for image interpretation. Furthermore, most of these satellites will offer in track *stereo* imagery. Combined with the very high resolution of the data, stereo imagery allows the direct generation of accurate *3-dimensional (3D) information* via standard photogrammetric or computer vision methods. Additionally, many systems can provide *multi-view angle* sensing; thus allowing to monitor particular parts of the earth's surface from different tracks of the satellite. In this way, $6-8$ images of the same part of the scene will be available. And finally, some sensors will even generate *hyperspectral* data involving 200 different spectral bands (at a spatial resolution of $8\,m$). Although the analysis of the individual bands *per se* already will be a tremendous task, many new applications are expected from the combined use of the bands. In fact, they can provide detailed *context* information for the interpretation of the very high resolution panchromatic imagery via *data fusion*.

A similar evolution in the availability and type of airborne imagery has resulted in a paradigm shift for aerial image interpretation during the past few years. Indeed, the increase in spatial resolution of the data from medium

---

* This work is partially supported by EU ESPRIT ltr 20.243 *"IMPACT"*.

scale (approx. $1\,m$) to high resolution (approx. $10\,cm$) has moved the focus of attention from larger scale applications such as road extraction and mapping to smaller scale applications such as the modelling and 3D reconstruction of buildings (cf. [3, 4]). Moreover, the interpretation of the data is increasingly being performed by *reasoning in 3D world space* (see e.g. [2, 9]) instead of generating 3D information from monocular analyses of the individual images as obtained by pattern recognition procedures. For example, in the case of building reconstruction most model-based approaches aim at identifying 3-dimensional building models (to various degrees of specification) from their predicted projection in the images or from a 3D reconstruction of the scene (see e.g. [4]). As a consequence, traditional stereo methods are being replaced by multi-view approaches which not only allow to improve the accuracy of the reconstruction, but also permit blunder correction due to the redundancy in the images and yield more complete reconstructions by the complementarity of the viewpoints.

In this paper, we investigate the feasibility of extracting 3-dimensional topographic information from multiple very high resolution satellite images. Section 2 briefly discusses a 3-dimensional, multi-view based method for automatic modelling and 3D reconstruction of buildings from aerial images. Section 3 then focusses on the effects of the lower resolution of satellite images on the performance of the method and proposes some extensions and alterations in order to operate with satellite data. The differences in accuracy between the 3D reconstructions obtained from aerial and satellite imagery is discussed as well. Finally, section 4 formulates the conclusions drawn from this analysis.

2. Automatic Modelling and 3D Reconstruction of Urban Buildings from Aerial Imagery

Buildings encountered in residential areas in European cities exhibit a wide variety in their shapes. Many roofs neither are flat nor are composed of simple rectangular shapes. This limits the use of pre-defined roof models for their extraction and reconstruction. In [1] it is therefore proposed to model a house roof as a collection of planar *polygonal patches*, each of which encloses a compact area with consistent photometric and chromatic properties, and that mutually adjoin along common boundaries. The strategy discussed here follows this philosophy by first constructing from the images. a *polyhedral model* of the roof structure, which captures the topology of the roof, but which might not be very accurate in a metric sense; then improving the *metric accuracy by fitting the model to the data*; and finally, the building model is completed by adding vertical walls. The modelling process is formulated as a feed-forward scheme, the main parts of which will be presented next. Figure 2.1 illustrates each part of the algorithm on a real example, taken from

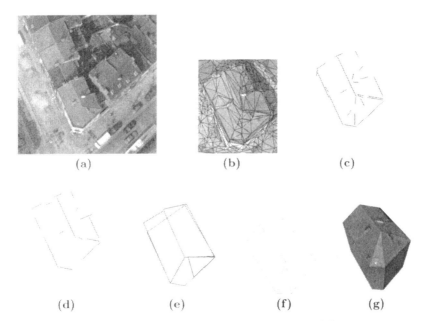

Fig. 2.1. The different steps of the reconstuction method: (a) A detail from an aerial image; (b) a part of the triangulation: (c) the edges contained in selected regions; (d) the 3D line segments reconstructed from 4 views; (e) the initial polygon hypotheses; (f) the polyhedral roof model; (g) the final reconstruction.

a state-of-the-art dataset produced by Eurosense Belfotop n.v. and consisting of 8 *cm* resolution colour images of residential areas in Brussels. The four-way image overlap guarantees that each building is visible in 4 to 6 images. Our method requires at least 3 views to be present. For a more detailed description of the method, the interested reader is referred to [7].

Building Selection and 3D Feature Generation. The reconstruction method uses straight line segments as the basic image features. Due to the nature of the scene, generally a large number of line segments is found in the images. For an efficient processing, it is desirable to divide the set of line segments into relatively small parts containing only one or a few buildings. As a house roof generally is constructed from the same roofing material, roof structures are likely to correpond to image regions with a fairly homogeneous colour distribution. Therefore, a triangulation network is constructed in the image(s) such that the extracted line segments coincide with edges of the triangles and that the colour content of a triangle's interior is fairly homogeneous. Image segmentation is then performed by selecting a triangle and growing a region from it by merging adjacent triangles that have the same mean colour vector as the selected one. These regions correspond quite well to actual roof surfaces and are quite stable over the different views. Therefore, all further processing

A 3-D Multi-View Strategy for Remotely Sensed Image Interpretation   151

is restricted to (one or a few) corresponding regions only. This allows us to keep the combinatorics under control at all stages of the algorithm.

The line segments contained in the selected regions are used for matching and possible reconstruction. As the regions are relatively small and fairly homogeneous, they will contain only a small number of line segments, mostly situated at the region boundaries. Matching line segments between corresponding regions in different images thus is rather easy. Most mismatches are ruled out by using the trifocal constraints that must hold between any three views of a stationary scene. For all line segments that are matched across 3 or more views, a 3D reconstruction is computed by bundle adjustment using the flight information and the calibration data of the camera.

Polygonal Patch Formation. Next, the reconstructed 3D line segments are grouped into coplanar configurations. Starting with the longest ones, 2 line segments are selected in a region. If the orthogonal distance between the 3D lines is small, a plane is constructed that fits to the line segments in a least-squares sense. In that case, the other line segments in the region are tested for coplanarity with the hypothesized plane. All line segments that satisfy this constraint are included in the defining set of the plane and the plane's equation is updated. This process is repeated until no more plane hypotheses can be formed from the selected region.

As the regions correspond well to roof structures, most plane hypotheses will correspond to planar patches of the roof. Thus, polygonal patch hypotheses can be formed directly from the line segments in the defining set of an hypothesized plane. An initial hypothesis is formed by constructing the convex hull of the line segments and adapting its boundary to incorporate line segments, one of whose end points contributes or is close to the hull. Due to the hull construction, 3D edges will be introduced at places where 3D information about the roof structure is missing. For example, no 3D line segments were generated for the rafters at the front of the corner house in figure 2.1 (d), because their edges have not been recovered well in the images. The convex hull computation has introduced these rafters when constructing the side patches (see figure 2.1 (e)). The position of the right front rafter, however, is not correct. So, before combining the polygons into a polyhedral roof model, these newly induced edges should be compared to the data, and erroneous edges must be corrected. To this end, it is first verified whether there are reconstructed 3D line segments that are close to the hypothesized edge and overlap with a significant part of it. If this is not the case, then the hypothesized edge is back-projected into each of the images, and one looks for supporting 2D line segments in at least one image. The error tolerances in any of the above constraints is very tight, in order not to rule out simple (but possibly correct) polygonal patches too early in the process. For the edge hypotheses that fail these consistency tests, alternative hypotheses, aiming at higher regularity (symmetry, parallellism, rectangular shapes, ...) for the polygon, are generated and verified by back-projection in the images.

Fig. 2.2. Two views of the 3D reconstruction of the houses in figure 2.1 (a).

By adopting a strategy of hypothesis generation in 3D world space and verification in the images we not only are capable of exploiting all the available image data at every step in the algorithm, but also to treat all views in an equal manner.

Roof Model Generation and Model Fitting. Roof modelling is completed by combining the polygonal patches into polyhedral roof models. For this purpose, a label is attached to each edge of the polygons, according to its 3D position in the polygon. In particular, non-horizontal lines are labelled *rafter*, and the horizontal ones are divided into *ridges* and *gutters*, depending on their height with respect to the average height of the polygon and the type of their neighbouring edges. Then the polygons are glued pairwise along edges bearing the same label and which nearly coincide. Glueing starts from the ridges, continues with the rafters, and ends with the gutters. Finally, all polygons that have not been glued, are removed.

During the 3D grouping and roof modelling stages of the algorithm, the metric accuracy of the reconstruction has been neglected. It is restored by fitting the roof model to the original image data. The roof model is converted into a *wireframe model* (figure 2.1 (f)), which is projected into all the images by using the flight information and the calibration data of the camera. The idea now is to adapt the 3D position of the vertices of the wireframe so as to bring the projected edges into as best as possible accordance with the observed image data. This is done by maximizing an *energy functional* which measures the average intensity gradient along the projected edges of the wireframe in each of the images together with the deviation of each vertex from its original 3D position. After optimization, the 3D building model is completed by adding vertical walls to the reconstructed roof. Figure 2.2 shows the result of applying the algorithm to all the houses in figure 2.1 (a). Figure 2.3 shows another example.

Fig. 2.3. Another aerial image *(left)* and the reconstructed house row *(right)*.

3. Operation with Satellite Imagery

To assess the potential of the methodology described in section 2 for the extraction of 3-dimensional information from very high resolution satellite imagery, it is investigated below how strongly the performance of the algorithm depends on the resolution of the images and which extensions and alterations are needed to let it operate with spaceborne imagery of 80 cm spatial resolution.

Lowering the Resolution of the Image Data. The success of the reconstruction method in section 2 depends on whether the correct polyhedral model of the roof structure can be constructed during the modelling phase of the algorithm. The latter relies on the accuracy and completeness of the initial polygon hypotheses that were formed from the reconstructed 3D line segments. As long as sufficiently many 3D line segments can be generated from the image data, the actual spatial resolution of the images does not really affect the roof modelling process. The first question therefore is whether *a spatial resolution of 80 cm suffices to extract the geometric image primitives* (2D line segments) *needed for the reconstruction*; and, more importantly, whether *these line segments can correctly be matched across the different views*, a necessary condition for their reconstruction in 3D world space.

To answer these questions, an experiment was carried out to determine the *minimal* spatial resolution needed for the method to work *without alterations*. A part of the test site, containing 32 buildings with different roof types (flat and non-flat, regularly and irregularly shaped), was selected, their images were progressively subsampled to systematically lower the resolution, and the algorithm was applied. It turns out that *the resolution may be lowered to 40 cm before a significant degradation in the performance of the algorithm is observed*. Moreover, the algorithm starts failing for resolutions lower than 40 cm, because then (most of) the extracted line segments contain less than 10 pixels, which is a critical length for stereo matching.

Figure 3.1 shows two reconstructions of two houses in figure 2.1 (a). The original (8 cm resolution) image detail and the corresponding 3D reconstruction is presented on the left; and the same is given on the right for the

154   T. Moons et al.

Fig. 3.1. *Left:* Original image and final reconstruction. *Right:* Subsampled image and final reconstruction.

subsampled (40 cm resolution) image. As can be seen, there is no qualitative difference between the two reconstructions. In fact, the maximum observed deviation in the world coordinates of the reconstructed roof vertices obtained from the original and from the subsampled images is 10 cm. This relatively small difference in metric accuracy of the final reconstruction – especially, when compared to the large reduction in spatial resolution – is due to the fact that the connectivity of the wireframe used for estimating the metric parameters of the reconstruction through the fitting process, globally compensates for the loss in location accuracy of the individual edges in the subsampled images.

Introducing an Additional Grouping Step. To investigate the performance of the algorithm on 80 cm resolution data, a second experiment was performed with images of industrial buildings. These buildings typically have a polygon shaped flat roof whose dimensions are significantly larger that those of residential buildings (see figure 3.2 (left)). Even at low resolutions, they generate relatively long straight line segments in the images; thus eliminating the difficulties reported above. As expected, most line segments of the roof structures can be matched correctly across three or more views and a sufficient number of 3D line segments is generated from these matches. But now the effect of the resolution on the accuracy of the 3D reconstruction comes into play: Height differences up to 4 m between individual, reconstructed 3D

Fig. 3.2. *Left:* Aerial image of an industrial building. *Middle:* Enlarged detail of the left view. *Right:* Enlarged corresponding detail of the subsampled image.

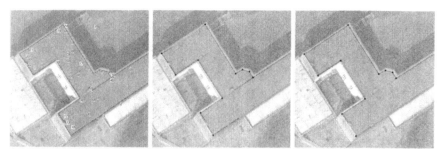

Fig. 3.3. Monocular grouping: extracted line segments *(left)*, detected corners *(middle)*, and constructed polygon *(right)*.

line segments of the same horizontal roof were observed. As a consequence, many plane hypotheses were formed by the algorithm, but only few of them contained sufficiently many 3D line segments to generate (partially) valid initial polygon hypotheses, thus causing the construction of a roof model to fail.

To overcome this problem, an additional step performing monocular grouping of the 2D line segments to extract the roof's outline in each image, is introduced before plane hypotheses are formed. Several monocular grouping methods have been proposed in the literature, mostly aiming at detecting rectangular structures (see e.g. [5, 6, 8]). Here, another strategy was chosen for two reasons: Firstly, the building shapes in our data set are largely non-convex; and secondly, since the processing is restricted to a few selected regions corresponding to the actual roof structure, only a relatively small number of line segments has to be grouped into a polygon. The strategy is illustrated in figure 3.3: For any two non-parallel line segments in a region, one of which having an end point that is close to the other line segment, the point of intersection of the underlying 2D lines is computed and a corner entry is created. This *corner* entry consists of the image coordinates of the intersection point, together with the direction and orientation of the half-lines containing (the largest part of) the 2D line segments that generate the corner. For every corner, a tree structure is build as follows: The vertices of the tree are corners; and an edge is constructed if the child vertex is coincident with the half-line of the parent that does not contain the grandparent. From this tree, the longest paths that can be formed, starting from both branches of the tree's defining corner, are selected and concatenated into a maximal chain of connected corners. Finally, the longest among all maximal chains of all corners, is selected as being (part of) the outline of the polygon. As most line segments belong to the boundary(s) of the selected region(s) in the images, and as these regions correspond well to roof surfaces, this longest chain of corners generally describes the outline of the roof in the image(s) quite well.

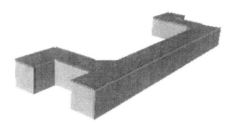

Fig. 3.4. 3D reconstruction of the building shown in figure 3.2.

When the outline of the roof is determined in every image, the original 2D line segments lying on these outlines are matched and reconstructed in 3D world space as before. Since these 3D line segments are supposed to lie in the roof plane of a flat-roofed building, the coplanar grouping step becomes superfluous, and the plane hypothesis is formed by fitting a plane to *all* reconstructed 3D line segments of the roof's outline. Then, an initial polygon hypothesis is formed and reconstruction continues as before. Observe that for updating the polygons one can profit from the extracted roof outlines in the individual images. Figure 3.4 shows the reconstruction of the building in figure 3.2, obtained in this manner.

Note that the 3D modelling phase in this extended algorithm still starts from the same 3D line segments, whose accuracy suffered from the low(er) resolution of the images, as before. But, because the plane hypothesis is now formed by using *all* available reconstructed 3D line segments of the roof plane, a much better accuracy for the height estimate of the roof is obtained. In our experiments, the maximum observed deviation in a roof's height between the final reconstructions obtained from the original and the subsampled images is $2.1\,m$.

The adapted version of the algorithm proved successful in reconstructing large, flat-roofed, industrial types of buildings with an acceptable accuracy from $80\,cm$ resolution imagery. Urban houses, however, cannot be handled in this way. The best one may expect to obtain from $80\,cm$ resolution data for dense urban areas is a 3D citymodel composed of polyhedra representing *blocks* of buildings and indicating the *average height* of the buildings in each block. An example is given in figure 3.5. Reconstruction succeeded because all houses have gable roofs which are very well aligned. In particular, the gutters and the ridges of the roofs in the images are quite co-linear and their common direction is dominant in the image part. This dominant direction is recovered automatically by means of a Hough transform [11]. Long straight line segments idealising the union of the gutters and the ridges of the houses were then formed by a co-linear grouping of the line segments having that dominant direction. To improve the accuracy of their location, these long line segments were fitted to the images, before stereo matching and 3D re-

Fig. 3.5. *Left:* Subsampled version of the left image in figure 2.3. *Right:* House row reconstructed from the subsampled images.

construction was started. Compared to the poor resolution of the image data (cf. figure 3.5 (left)), the 3D reconstruction is quite impressive. Moreover, the deviation in the average height of the ridges in the original reconstruction in figure 2.3 (right) and the height of the ridge in the reconstructed building block of figure 3.5 (right) is 31 cm. It must be observed, however, that both the building block and the used method are very specific and tuned to this example. Hence, in contradistinction to the results obtained for large, flat-roofed buildings, this building block example can be considered only as an indication of what might be possible for dense urban areas with 80 cm resolution data. Generalisation of this result will only be possible if one succeeds in developing generic methods for large scale structure detection in monocular images of dense urban areas. Maps may be a helpful tool in this respect [10].

4. Conclusion

The aim of this paper is to assess the feasibility of extracting 3-dimensional information about man-made objects from new generation satellite imagery. The very high resolution of the image data permits the use of geometric image features for image interpretation and facilitates the direct generation of 3-dimensional data from in track stereo pairs. Multiple views from different viewpoints of the same part of the scene, obtained through multi-view angle technology, adds redundancy for blunder detection and correction. Under these conditions, a 3-dimensional, multi-view based paradigm for image interpretation – as used for aerial images – can be adopted. The underlying philosophy is to generate, as soon as possible, from the images reliable geometric 3D features, exploiting the multi-view geometric constraints for blunder correction. Scene analysis is then performed by reasoning in 3D world space and verified in the images by means of a hypothesis generation and verification procedure in which the decisions are taken on the basis of a multi-view consensus.

158   T. Moons *et al.*

As an example of such an approach, a method for automatic modelling and 3D reconstruction of urban buildings from aerial images is discussed and the effect of the lower resolution of satellite data on the performance of the algorithm has been investigated. It turns out that the spatial resolution of the images may be lowered to 40 cm before a significant degradation in the performance of the algorithm is observed. Above the 40 cm border, no significant loss of metric accuracy in the final reconstruction is observed. The reason for failure below the 40 cm border is that most of the extracted straight line segments are too small to allow a correct matching over different views. For images with a spatial resolution of 80 cm the algorithm can be adapted to generate acceptable results for the 3D reconstruction of large, industrial types of buildings. In fact, only an additional monocular grouping step is introduced to extract the outline of the roofs in the individual images in order to facilitate the polygonal patch formation in 3D world space. The lower resolution also affects the metric accuracy of the final reconstruction: A maximum deviation of 2.1 m has been observed. Finally, it is noted that for dense urban areas one might expect to generate 3D models composed of polyhedra representing blocks of buildings from 80 cm resolution imagery, and an estimate of the average height of the buildings in that block, provided one succeeds in developing generic methods for large scale structure detection in monocular images of this type of area.

References

1. F. Bignone, O. Henricsson, P. Fua, M. Stricker, "Automatic extraction of generic house roofs from high resolution aerial imagery", B. Buxton and R. Cipolla (eds.), *Computer Vision – ECCV'96*, LNCS 1064, pp. 85–96, Springer-Verlag, Berlin, 1996.
2. A. Fischer, T. H. Kolbe and F. Lang, "Integration of 2D and 3D reasoning for building reconstruction using a generic hierarchical model", in *Semantic Modelling for the Acquisition of Topographic Information from Images and Maps*, W. Förstner and L. Plümer (eds.), pp. 159–180, Birkhäuser-Verlag, Basel, 1997.
3. A. Grün, O. Kübler and P. Agouris, *Automatic Extraction of Man-Made Objects from Aerial and Space Images*, Birkhäuser-Verlag, Basel, 1995.
4. A. Grün, E. P. Baltsavias and O. Henricsson, *Automatic Extraction of Man-Made Objects from Aerial and Space Images (II)*, Birkhäuser-Verlag, Basel, 1997.
5. J. Mc Glone and J. Shuffelt, "Projective and object space geometry for monocular building extraction", in *Proceedings IEEE Conference on Computer Vision and Pattern Recognition (CVPR '94)*, Seattle, WA, pp. 54–61, June 1994.
6. R. Mohan and R. Nevatia, "Perceptual organisation for scene segmentation and description", *IEEE Transactions on Pattern Analysis and Machine Intelligence*, vol. 14, no. 6, pp. 616–635, 1992.
7. T. Moons, D. Frère, J. Vandekerckhove, and L. Van Gool, "Automatic Modelling and 3D Reconstruction of Urban House Roofs from High Resolution Aerial Imagery", in *Computer Vision – ECCV'98*, H. Burkhardt and B. Neumann (eds.), LNCS 1406, pp. I.140–I.425, Springer-Verlag, Berlin, 1998.

A 3-D Multi-View Strategy for Remotely Sensed Image Interpretation   159

8. R. Nevatia, C. Lin and A. Huertas, "A system for building detection from aerial images", in *Automatic Extraction of Man-Made Objects from Aerial and Space Images (II)*, A. Grün, E. P. Baltsavias and O. Henricsson, (eds.), Birkhäuser-Verlag, Basel, pp. 77–86, 1997.
9. M. Roux and D. Mc Keown, "Feature matching for building extraction from multiple views", in *Proceedings ARPA Image Understanding Workshop (IUW'94)*, Monterey, CA, pp. 331–349, 1994.
10. M. Roux and H. Maître, "Three-dimensional description of dense urban areas using maps and aerial images",in *Automatic Extraction of Man-Made Objects from Aerial and Space Images (II)*, A. Grün, E. P. Baltsavias and O. Henricsson, (eds.), Birkhäuser-Verlag, Basel, pp. 311–322, 1997.
11. T. Tuytelaars, L. Van Gool, M. Proesmans, and T. Moons, "A cascaded Hough transform as an aid in aerial image interpretation", in *Proceedings International Conference on Computer Vision (ICCV'98)*, Bombay, India, pp. 67–72, January 1998.

3D Exploitation of SAR Images

Regine Bolter and Axel Pinz

Computer Graphics and Vision, Graz University of Technology,
Münzgrabenstraße 11, A-8010 Graz, Austria.
e-mail: {pinz,bolter}@icg.tu-graz.ac.at

Summary. The special nature of synthetic aperture radar (SAR) images as opposed to optical images requires highly specialized processing. This paper is devoted to the task of 3D surface reconstruction from multiple radar images. We discuss SAR-specific modifications of well known algorithms (stereo, shape-from-shading), as well as new algorithms incorporating knowledge about the SAR image formation process. Finally, we present a new method to integrate information from several sensors, modalities, or algorithms and discuss its potential applications in SAR data processing.

1. Introduction

It is well known [8] that many kinds of significant geometric distortions and defects occur in radar images which are usually more difficult to handle compared to optical imagery. In mountainous terrain, the occurrence of layover regions is possible: On steep slopes, the conversion of the side-looking slant-range radar geometry to ground-range is ambiguous, so that image pixels cannot be associated with a specific location on the ground. However, since this effect depends on the shape of the terrain, it can be predicted where the relief is known. The same applies to radar shadows, another adverse peculiarity occurring on steep backslopes.

In this paper we discuss several image analysis tools especially tuned for SAR images: SAR image simulation, automated matching, and shape-from-shading. We then present in detail a SAR stereo analysis experiment in layover areas. Future work will concentrate on the integration of several methods based on the new paradigm of 'active fusion'.

2. Test Site

Our principal study areas are the Ötztal test site, a rugged terrain on the main ridge of the Austrian Alps and the Maxwell Montes area on the Planet Venus, also a highly mountainous terrain.

The Ötztal test site is characterized by strong topographic relief, with elevations ranging from 1750 m to 3750 m. The surface cover consists mainly of alpine vegetation, rocks, snow and glaciers.

We used test images of the Ötztal area acquired by the ERS-1, X-SAR, and J-ERS-1 sensors. The most pronounced layover can be found in the

ERS-1 images, due to the steep look angle of 23°. From the same area a high resolution, high accuracy Digital Elevation Model (DEM) with a grid width of 25 m was also available.

The SAR images of planet Venus were acquired by NASA's Magellan probe in 1990–1994. More than 95% of the Venusian surface were imaged by the on-board radar sensor, resulting in over 400 Gbytes of SAR imaging data. One of the major goals of the mission was the computation of a high-resolution map of the whole planet, a tool which plays a crucial part in the geophysical analysis of all planetary processes.

Magellan images were acquired in three cycles, denoted as Cycles I, II and III. In Cycles I and III, the radar was looking to the left (East, as the satellite descended over the North pole), whereas in Cycle II the imaging configuration was right-looking. This means that for many areas on Venus a same-side stereo pair as well as a corresponding opposite-side image are available. However, in comparison to the Ötztal test site, there exists only a low resolution, low accuracy DEM from Venus, resulting from Magellan altimeter measurements with a resolution of 1 km.

3. Image Analysis Tools

For detailed studies of geometric effects, as layover and shadow, we have developed various image analysis tools especially suited to SAR images. This includes matching algorithms for stereo intersection, a shape-from-shading algorithm for DEM refinement, and SAR simulation.

3.1 Simulation

The development of new image processing and feature extraction algorithms broadens the need for the application of simulated images, since they provide inexpensive and flexible test material. Through the lack of ground truth data on Venus, simulation is the only tool to verify results obtained from Magellan SAR images.

The main design goal of our simulator was to achieve high geometric accuracy. A thorough description of our simulator can be found in [5]. The simulation is based on a digital elevation model, the precise modeling of the sensor flight path and a rigorous mapping model based on SAR range/Doppler equations ('parametric mapping model').

A comparison of corresponding sections of a simulated SAR image (left) and the actual SAR data acquired by ESA's ERS-1 satellite from an ascending orbit (right) can be seen in figure 3.1. The bright structure in the middle of both pictures is a layover configuration. Besides the absence of speckle noise in the simulated image, the two images correspond highly, especially in the layover area.

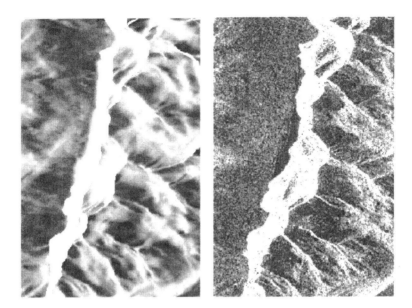

Fig. 3.1. Simulated SAR image (left) and the corresponding section of the original ERS-1 image from the Ötztal test site. The area covers approximately $4 \times 6\ km^2$.(©ESA).

When dealing with steep slopes in the terrain, the computation of the so called 'layover and shadow map' becomes an important requirement. This map, which can be computed in both image and DEM geometry, contains information on the location of layover and shadow areas. One possible application of such a map is to mask those areas of an image where the radiometric information is not reliable.

In mountainous regions, radar backscatter predominantly depends on local incidence angle, as opposed to terrain cover. Therefore we have employed a backscatter function which models the effects of several classes of terrain cover often found in mountainous areas. The effects of speckle noise to SAR images were also modelled appropriately.

3.2 Automated Matching

Another important image analysis tool for the 3D exploitation of SAR images is automated matching. Applied to radar images, the matching algorithm employed has to overcome considerable noise effects as well as severe radiometric and geometric discrepancies. We obtained best results from a correlation algorithm working on a hierarchical basis, considering also distinct features in the vicinity of a match point [4]. Promising results were also obtained with a new developed adaptive local description matching algorithm [11].

3.3 Shape-from-Shading

Shape-from-shading offers the advantage to reach high resolutions of the reconstructed digital elevation models in areas where dense disparity maps can not be expected from stereo matching. This is the reason why the algorithm is mainly used to refine previously known elevation models of low resolution. In contrast to other reconstruction methods shape-from-shading estimates local surface slopes instead of absolute height values. From the grey values of the pixels, conclusions on the normal vectors of the local surface patches can be drawn. Shape-from-shading requires the knowledge of scene illumination and of the reflection properties of the surface.

In our approach shape-from-shading is formulated as an optimization problem. A cost function, which balances the roughness of the estimated elevation model and its consistency with the SAR images, is to be minimized. The implemented iterative algorithm simulates the two SAR images. Then the differences between real and simulated images are evaluated and a new set of local surface slope estimates is calculated. The elevation model is then transformed to frequency domain, since the Fourier representation allows some operations, like filtering and enforcement of integrability of the slopes, to be performed more easily. Information from the previously known elevation models is also transformed in the frequency domain. The simulation of the images uses physical models of reflectance behavior as well as functions determined from measurements.

The implemented algorithm was tested on images acquired by NASA's Magellan space probe and on images of a test area in Djibouti, East Africa, acquired by XSAR and SIR-C. The input images and results can be seen in figure 3.2. The area covers approximately 40.96 km^2 and is located at 42° E, 12° N. The resolution of both SAR images is 12.5 m per pixel and both SAR images were captured from the left, the XSAR image (a) with a look angle of 34.8°, the SIR-C image (b) with a look angle of 35.7°. The resolution of the original DEM (c) is much coarser, about 200 m per pixel. The digital elevation model refined by shape-from-shading can be seen figure 3.2(d). Note the increase of detail in the refined DEM, for example the appearance of new structures on the middle left side of the refined DEM. These structures are visible in both SAR images, but not in the coarse DEM.

4. SAR Stereo Analysis Experiment in Layover Areas

In an experiment on three overlapping Magellan images of the Venusian surface we show how the special consideration of a possible layover situation can improve the terrain model derived by stereo methods [1]. The implemented semi-automated procedures reconstruct terrain slopes and at the same time are able to distinguish between foreshortening and layover. Simulation is used

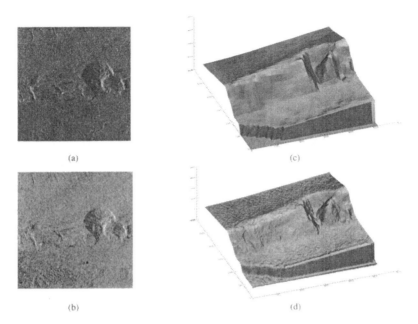

Fig. 3.2. Shape-from-shading experiment in Djibouti: (a) XSAR image, (b) SIR-C image, (c) original low resolution DEM, (d) DEM refined by shape-from-shading.

as a tool for quality control and to merge information from opposite-side SAR images.

In Magellan SAR data a considerable amount of layover can be found due to the steep look angle used. Figure 4.1 shows three Magellan SAR F-BIDRs (Full- resolution Basic Image Data Records) of our test site on Venus. The images were acquired during Cycles I (a) and III (b), both left looking with look angles of 33.5° and 17.5°, respectively, and during the right looking Cycle II (c) at 25.0°. Images (a) and (b) show salient bright stripes in the middle of the images, which are either foreshortening or layover. The corresponding structure in image (c) is rather dark. Our investigations concentrate on these stripe structures.

Our approach is motivated by a study on the manual extraction of height information from Magellan data [2]. In that work, a method to estimate local height differences and corresponding slopes in layover and shadow areas measuring the width of the layover or shadow area is developed. The method is restricted to discretely dipping surfaces.

Initially standard stereo analysis techniques [6] were applied to the test site. Simulation with the resulting DEM was used for verification. The regions outside the layover area were sufficiently reconstructed, the layover area itself however, did not arise in the simulated image.

The analysis showed, that the appearance of layover is closely related to the appearance of foreshortening. Due to the different look angles in the

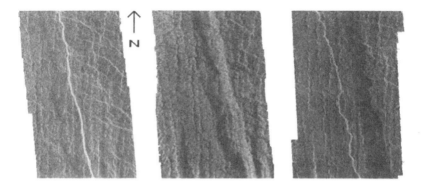

Fig. 4.1. F-BIDRs captured during Orbits 1202, 4793, and 3029. The area is approximately 24 × 38 km^2 and located on Venus' surface at 29.5° S, 142.5° E.

stereo image pairs, three possible cases exist: layover in both images, foreshortening in both images or layover in the image with the steeper look angle and foreshortening in the other image.

Layover and/or foreshortening areas in the SAR stereo image pair are segmented applying local adaptive threshold techniques. New matchpoints at the edges of the segmentation are computed and integrated into existing matchpoints derived from a standard matching algorithm [4].

Stereo analysis using the refined locations yields two solutions for the digital elevation model and values for the resulting slopes and height differences. For each solution a simulated opposite side image is generated. Comparing the two simulations to the real opposite side image, it can be established, if layover actually occurred in one, both or neither of the stereo images. For the test site in figure 4.1 layover occurred in both images. The resulting digital elevation model can be seen from figure 4.2.

Our analysis showed, that improvements in the stereo processing of layover areas are in principle possible. The results of the automated procedure differ only slightly from the manual measurements Connors [2] made. The estimated errors are comparable to the measurements in [2]. If no reference digital elevation model is available, simulation is the only tool to verify the results. A future study might focus on the refinement of the segmentation algorithm (e.g., based on edge or texture information), as well as the possibility of adding surface detail in the reconstructed layover areas by applying shape-from-shading to the opposite-side image.

5. Active Fusion

It is apparent from th eprevious sections that in SAR image processing – as in any other remote sensing application – we are dealing with *multisource*

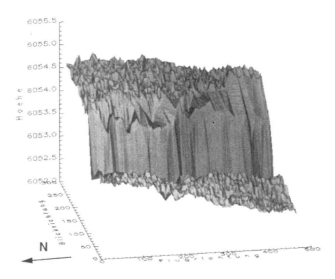

Fig. 4.2. Digital elevation model resulting from the stereo images in figure 4.1.

situations. In general more than one image, same or opposite side, or simulations is required, possibly integrating multitemporal or multisensor data, and several algorithms may have to be combined, e.g. stereo and shape-from-shading. Specifically, multisource data integration in remote sensing requires solutions for the following cases:

- Spatial registration: In order to compare multisource images at the pixel level, geocoding is required. Registration requirements get less severe at higher levels of abstraction (pixel level → feature level → symbol level). At the symbol level, often also called 'decision level' [3], no spatial registration is required at all.
- Dealing with uncertainty: Different kinds of 'ignorance' have to be treated. Many categorizations have been tried (e.g., [7]). Important categories are: Vagueness, ambiguity, incompleteness, conflict. Different mathematical frameworks have been used to model uncertainty: Probability theory, Dempster Shafer theory of evidence, fuzzy set and fuzzy measure theory.

We have developed a new method termed *'active fusion'* and tested it with several sample applications including a simple remote sensing task [10]. The basic concept of active fusion is as follows (figure 5.1):

1. A query defines the current task. Proper initialization of the system is assumed. This includes the generation of a problem description with representation of current state and goal state, and the knowledge of available data sources and/or actions.
2. The current state is assessed. Are there still unused sources or new actions to try? Can improvements of the result be expected? Is the cur-

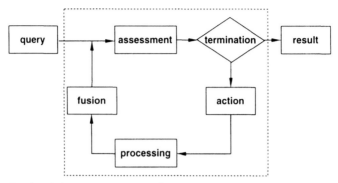

Fig. 5.1. Top level decision – action – fusion loop in an active fusion system with sequential processing and integration of information sources.

rent result already satisfactory with respect to the query? The remaining datasources/actions are ranked based on their expected utility.
3. If the current state has been assessed to deliver a satisfactory result, the system terminates.
4. The most promising action is performed next . This may be the use of a new data set, a different processing method (often termed 'soft sensor') of an already used dataset, or an interaction with the environment. Imagine a mobile robot performing a general active vision task. A remote sensing example might be the on-line tasking of a next generation satellite.
5. New data obtained by the previous action are processed. Any level of abstraction (sensor, pixel, feature, symbol) is possible.
6. Fusion: The current state of the system is updated based on the information obtained by the previous processing step.

In comparison to standard multisource data integration methods in remote sensing, active fusion is a very general concept of *control* in computer vision [9]. A decision is required based on multiple information sources tainted with different kinds of uncertainty. The situation has to be *evaluated*, and a *decision* has to be taken. *Entropy measures* are used to evaluate the current situation. Let $\mathbf{p} = \{p_i(\mathbf{x}), i = 1, \ldots, N\}$ be a probability distribution denoting the probability, that given a current situation $\mathbf{x}$, a classifier would vote for class i. Then,

$$H(\mathbf{p}) = -\sum_{i=1}^{N} p_i \log_2 p_i$$

denotes Shannon entropy $H(\mathbf{p})$, which approaches zero for a perfectly sharp decision ($p_k = 1, p_i = 0, i \neq k$). This leads directly to a termination criterion which demands $H < threshold$. As long as the termination criterion is not fulfilled, a decision is required, which information source should be used next. There are many possible ways of estimating the expected utility of a source. All of them require learning and application dependent modeling (in [10],

Bayesian networks to model a multisource, multitemporal remote sensing situation were used).

Pinz *al.* [10] describe a simple application of classifying crop categories of different agricultural fields based on a multitemporal Landsat dataset. In this work, active fusion was not applied to SAR problems yet. There are, however, many potential applications in the context of the methods for 3D exploitation of SAR images presented:

- Use simulated images to assess radiometric reliability of real images;
- Use shadow/layover information to guide processing/switch between processing modes;
- Fuse the results of different stereo matching algorithms;
- Directly fuse stereo and shape-from-shading based on confidence measures.

6. Conclusions

SAR image analysis requires specialized algorithms and tools and is more fragile than optical image analysis. We have presented an overview of a bundle of methods developed to generate 3D surface descriptions from SAR images. In an experiment on the reconstruction of a layover distorted surface slope from three overlapping SAR images we demonstrated the application of two of the methods, automated matching and simulation. In a future study shape-from-shading applied to the opposite-side image might be used to add surface detail in the reconstructed area.

Furthermore, the idea of active fusion as a new means of control (selection of input data as well as selection and tuning of processing algorithms) has been introduced and its potential applications to SAR processing has been discussed.

Acknowledgement. We gratefully acknowledge support of our work by the following contracts: FWF (Austrian science foundation) contract No. S7003, task 3.1; Austrian ministry of science contracts GZ 601.560/2–II/6/94 and GZ 601.574/2–IV/B/9/96.

References

1. R. Bolter, M. Gelautz, and F. Leberl, "Simulation based SAR-stereo analysis in layover areas", in *Proceedings of the International Geoscience and Remote Sensing Symposium, IGARSS'98*, (Seattle, WA), 6–10 July 1998, vol. I, pp. 345–347, 1998.
2. C. Connors, "Determining heights and slopes of fault scarps and other surfaces on venus using Magellan stereo radar", *Journal of Geophysical Research*, vol. 100, no. E7, pp. 14361–14381, July 1995.

3. Belur V. Dasarathy, *Decision Fusion*. IEEE Computer Society Press, 1994.
4. R. Frankot, S. Hensley, and S. Shafer, "Noise resistant estimation techniques for SAR image registration and stereo matching", in *Proceedings of the International Geoscience and Remote Sensing Symposium, IGARSS'94*, (Pasadena, CA), 8–12 August 1994, vol. II, pp. 1151–1153, 1994.
5. M. Gelautz, H. Frick, J. Raggam, J. Burgstaller, and F. Leberl, "SAR image simulation and analysis of alpine terrain", *ISPRS Journal of Photogrammetry and Remote Sensing*, vol. 53, pp. 17–38, 1998.
6. S. Hensley and S. Shafer, "Automatic DEM generation using Magellan stereo data", in *Proceedings of the International Geoscience and Remote Sensing Symposium, IGARSS'94*, (Pasadena, CA), 8–12 August 1994, vol. III, pp. 1470–1472, 1994.
7. G. J. Klir and T. A. Folger, *Fuzzy Sets, Uncertainty, and Information*, Prentice Hall, 1988.
8. F. Leberl, *Radargrammetric Image Processing*. Norwood, MA:Artech House, 1990.
9. A. Pinz, "Active fusion - a new concept for control in computer vision", in *Proceedings GDR-PRC ISIS, CNRS, Paris*, pp. 120–131, 1996.
10. A. Pinz, M. Prantl, H. Ganster, and H. Kopp-Borotschnig, "Active fusion - a new method applied to remote sensing image interpretation", *Pattern Recognition Letters*, vol. 17, pp. 1349–1359, November 1996.
11. S. Scherer, W. Andexer, and A. Pinz, "Robust adaptive window matching by homogeneity constraint and integration of descriptions", in *Proceedings of ICPR*, vol. I, pp. 777–779, 1998.

Visualizing Remotely Sensed Depth Maps using Voxels

Nilo Stolte

School of Computer Science and Electronic Systems, Kingston University,
Penrhyn Road, Kingston upon Thames, KT1 2EE, UK.
e-mail: stolte@dcs.kingston.ac.uk

Summary. This article presents a new method for visualizing remotely sensed depth maps by using voxels. Voxels have been introduced in the 1980's to accelerate Ray Tracing, a very time consuming rendering algorithm. With the advent of new technologies for obtaining volumetric data in medical images, voxels were immediately accepted for their representation. New kinds of algorithms were created for visualizing this kind of data. This new research area opened the door for completely new concepts and the idea of discrete graphics has flourished. However, the huge amounts of memory as well as high processing times generally necessary for accomplishing discrete graphics have practically discarded it as a practical visualization tool. New techniques to represent high-resolution volumes using hierarchical data structures and visualization techniques able to efficiently skip over empty space significantly changed this panorama. This is the ideal arena for normal computer graphics surfaces to be visualized in the voxel format. The visualization of remotely sensed depth maps can also benefit of these techniques and profit of a number of advantages over traditional graphics.

1. Introduction

Voxels are 3D versions of pixels and have been used fairly commonly by the computer graphics community. They have been used mainly for accelerating time consuming visualization algorithms such as ray-tracing and radiosity [3, 4, 5, 6, 11]. A new application for the use of voxels has arrived when 3D medical imagery such as Computed Tomography (CT) and Magnetic Resonance Imaging (MRI) started to appear. These data can be easily obtained in the voxel format. A whole new field in visualization then started to consolidate itself, namely "Volume Rendering" (also known as "Volume Visualization") [10]. The main problem in this area are the high rendering times necessary to render volumes, basically based on the ray-casting algorithm. In between standard computer graphics and Volume Rendering another technique called "Discrete Ray-Tracing" showed up [15]. It basically works as Ray-tracing but entirely in a voxel volume. Recent techniques [12] have been introduced to accelerate the rendering time of discrete scenes mostly containing empty space. This is the ideal arena for normal computer graphics primitives to be visualized in the voxel format. This requires a conversion from analytic surfaces to voxels,a process called "voxelization" [1, 8, 9]. Several kinds of surfaces can be converted to voxels, including one of the most modern modelling paradigm: implicit surfaces [2, 7, 13, 14].

Remotely sensed depth fields are basically terrains, that can be seen as surfaces. This kind of data can be immediately mapped into voxels and it can be easily visualized using the techniques explained in the former paragraph. The big advantage of using voxels in this context is that each "pixel" from the remotely sensed depth maps can be stored directly in each voxel. This eliminates the problem of texture mapping. Also clipping parts of the terrain to add new features is not necessary; the voxels occupying the area corresponding to new features can be simply deleted and the voxels of the new features superposed. New features can be all sort of objects, such as synthetic objects (i.e. objects that have been voxelized).

The use of voxels to represent remotely sensing depth maps is much more natural and makes manipulation of the data as simple as drawing objects on the screen. New techniques to visualize these kind of voxels models (surface-based voxel models) interactively, using GL/OpenGL points, enables stereoscopic visualization and manipulation using virtual reality techniques.

2. Voxels Storage: Octree

High resolution 3D grids are essential for good quality representation and rendering in discrete graphics. The same is valid for 2D screens. The image quality and the representation will be better in higher resolutions than in low resolutions. A few years ago, it was very uncommon to find high resolution frame-buffers in an ordinary computer. Nowadays, this situation has changed because the price of the memory has drastically dropped.

However, three-dimensional frame buffers consume much more memory than 2D frame buffers. For example, a 1024^3 3D grid is equivalent to 1024×1024^2 frame buffers. Therefore, the memory consumption for a 3D grid can be thought of as 3 orders of magnitude greater than an equivalent 2D buffer. Clearly, ordinary computers can still not support 3D grids of such high resolutions. Thus, it is not unusual to find fully or partially compressed volumes to compensate the enormous memory requirements. This problem resembles that of real time video image sequences, which require some kind of compression. Compression algorithms are very time consuming and for real-time video they are generally implemented in hardware. Unfortunately, there is no such available hardware to enable high-resolution discrete graphics in today's machines.

In 2D images only the colour and/or intensity stored in the pixel is enough to precisely describe one image. Three-dimensional data sets also generally store the intensity/colour in each voxel. However, this is not enough to completely describe the 3D volume for rendering purposes. In the rendering process the illumination is calculated using the normal vectors.

Equipments generating medical data sets generally deliver convoluted volumes. In these conditions, it is easy to approximate the normal vectors using a technique called *central difference*. In this technique, the voxel intensities

172   N. Stolte

are considered as implicit function values and the normal is calculated by applying the gradient to the function. Since the function is not known, the gradient is calculated by applying the partial derivatives definition. Nevertheless, the derivatives calculated in this way are correct only when the limit of the size of the voxel tends to zero. This would imply an infinite resolution volume. Obviously this is not possible since the voxel size is always greater than zero because of the finite resolution of the volume. Therefore, the normal vectors calculated in that way will be always approximate.

Another difficulty is that the normal vectors must be normalized in order to be useful in the rendering process. This demands an additional amount of time in the rendering process or extra memory to store the normalized normal vector in the voxels. Thus, the already very high memory requirements result completely unpractical if the normal vectors are stored in the volume.

Our solution is to represent in high resolutions only the part of the volume where the surface transitions take place. This can be accomplished by using an octree. In fact, the octree allows in this way to compress the volume, assuming that most of it is empty, which is generally the case in most common scenes. An interesting feature, though, is that the interior part of the volumes in this representation is known as opposed to polygon representations. Polygon representations (polytopes) do no allow the representation of their interior space. This can be seen as a significant disadvantage of polytopes.

The compression obtained with the octree allows the normal vectors storage in the voxels and at the same time very high resolutions in normal workstations.

Figure 2.1 illustrates the octree implementation representing an octree with 5 levels (diagram on the left side) and its visualization in the space (on the right). In practice, an octree with 9 levels is usually used, thus defining a volume of 512^3. In figure 2.1 five cells, one for each different level, are shown. Each cell has eight elements, all representing the eight equally sized subdivisions of the cell. At the last level (level k, such that $k = 4$), each element corresponds to a voxel of a 3D grid representing a $2^5 \times 2^5 \times 2^5$ volume. These voxels can be noted by their coordinates, (X, Y, Z), as follows: $V_0(30, 30, 30)$, – the only one not visible in the 3D representation – $V_1(31, 30, 30), V_2(30, 31, 30), V_3(31, 31, 30), V_4(30, 30, 31)$, – indicated in the figure, – $V_5(31, 30, 31), V_6(30, 31, 31), V_7(31, 31, 31)$.

The order of the voxels in a cell is the same as adopted in [4]. The definition of this order for an octree of n levels is given as follows. This octree represents a volume of $2^n \times 2^n \times 2^n$ resolution. Each coordinate X, Y, Z of a voxel in this volume is a binary number defined by an accumulation of powers of two. Let k be an index of a bit in the binary coordinate (where $k = n - 1$ is the rightmost bit, and $k = 0$ is the leftmost bit). Then, the index of an element in a cell is given by the following formula:

$$i_k = x_k + 2 \cdot y_k + 4 \cdot z_k,$$

where,

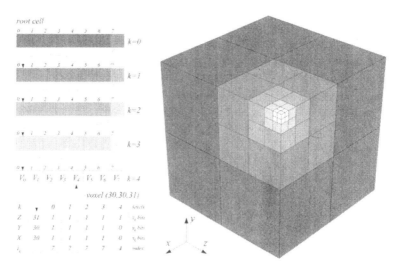

Fig. 2.1. Left: Octree in the memory; Right: correspondent volume in the space.

$$X = \sum_{k=0}^{n-1} x_k \cdot 2^{(n-1)-k} \qquad 0 \leq X \leq 2^n - 1$$

$$Y = \sum_{k=0}^{n-1} y_k \cdot 2^{(n-1)-k} \qquad 0 \leq Y \leq 2^n - 1$$

$$Z = \sum_{k=0}^{n-1} z_k \cdot 2^{(n-1)-k} \qquad 0 \leq Z \leq 2^n - 1$$

and, $x_k, y_k, z_k \in \{0, 1\}$.

It is clear by this definition the relationship between the voxels' binary coordinates and the levels of the octree. Each bit correspond to one octree level and the combination of the three bits of each coordinate in a level k forms the index of the element in each cell of the octree at the level k. This relationship is fully exploited in the implementation of the octree generation presented in the next section.

3. Octree Generation

Our octree is a classical pointer octree, where the root node is defined by a pointer called "octree", as shown in figure 2.2. This pointer points to an array of pointers with eight elements, each one representing one eighth of the original volume. Each of these arrays is called a cell. A null pointer means that the region is empty, while a non-null pointer points to another array of

```
char *octree;/* pointer to the first free octree byte */
char *free_space; /* pointer to the first free byte in a block */
int free_bytes; /* number of remaining free bytes in a block */
int X_ant, Y_ant, Z_ant, mask1, mask2;
init_octree() {
/* Initialize masks as follows (each square is a bit) */
/* n = number of octree levels, nb+1 = number of variable bits */
```

$mask1 \leftarrow$	nb		n+3	n+2	n+1	n	n-1	n-2		5	4	3	2	1	0
	0	...	0	0	0	1	0	0	...	0	0	0	0	0	0
$mask2 \leftarrow$	1	...	1	1	1	0	0	0	...	0	0	0	0	0	0

```
octree←free_space←alloc_block(); /* allocates one block */
free_bytes←Size_of_Block−Bytes_in_Cell;
free_space←free_space+Bytes_in_Cell;
push(octree);
/*variables to find common parent */
X_ant←0; Y_ant←0; Z_ant←0;
}
store_in_octree(X,Y,Z,input)
int X,Y,Z;
any input;
  { char **pcel;
    /* Ascend octree to find a common parent */
    while ( ( ( (X and mask2) ≠ (X_ant and mask2) ) or
              ( (Y and mask2) ≠ (Y_ant and mask2) ) or
              ( (Z and mask2) ≠ (Z_ant and mask2) ) )
      { pop();
        mask1←mask1<<1; mask2←mask2<<1;
      }
    pcel←pop();
    while (TRUE) /* Descends octree until the voxel*/
      { push(pcel);
        if (Z and mask1) pcel←pcel+4;
        if (Y and mask1) pcel←pcel+2;
        if (X and mask1) pcel←pcel+1;
        if ((mask1 and 1) = 0)
          { mask1←mask1>>1; mask2←mask2>>1;
            if (*pcel = 0) /* if node doesn't exist, creates it */
              { if (free_bytes<Bytes_in_Cell)
                  { free_space←alloc_block(); /* allocates */
                    free_bytes←Size_of_Block; /* one block */
                  }
                *pcel←free_space; /*creates and descends */
                pcel←free_space;
                free_space←free_space+Bytes_in_Cell;
                free_bytes=free_bytes-Bytes_in_Cell;
              }
            else pcel←*pcel; /* Otherwise descends only*/
          }
        else break; /* Leaf reached. Exit loop */
      }
    X_ant←X; Y_ant←Y; Z_ant←Z;
    *pcel←input;
}
```

Fig. 2.2. Octree generation algorithm.

eight pointers, further subdividing the region. This process continues until the leaf node is found, where each non-null pointer points to a voxel.

The efficiency of our octree lies into its simplicity. We keep one integer variable *"mask1"* with a set bit exactly at the bit position *"n"*, where *"n"* is the current octree level, which is the total number of octree levels in the beginning (see figure 2.2). We use this bit to filter the coordinates bits and to control the algorithm as in the octree ray traversal algorithm in [12].

The algorithm in figure 2.2 is given in a "C-like" pseudo-code. For the sake of clarity the type castings are omitted; each assignment is given by a $\leftarrow$, the logical commands are written with their names (and and or) instead of symbolically, and the recursive stack operations are denoted by push (to put an element into the stack) and pop (to remove an element from the stack).

Once initialized, the octree is dynamically created by calling the routine store_in_octree() for each new produced voxel. This function receives four parameters, the three voxel coordinates (X, Y and Z) and a pointer to the voxel content (*input*). In our case, it is the pointer to the surface normal in the voxel.

A significant feature of this algorithm is that it does not require descending all octree levels from the root. It starts from the *cell* where the last voxel was stored. In most cases the current voxel will lie in the same *cell* or in a nearby relative *cell*. If it does not lie in the same *cell*, the algorithm ascends some levels until the common parent is found. This happens in the first part of the algorithm.

To find the common parent we use the variable *mask2* as shown in the algorithm. This part is considerably efficient because it is translated to very few machine instructions and the variables used are always in the cache memory. The variable *mask2* is used to filter the most significant bits from the coordinate values. While the most significant bits of the current voxel coordinates filtered by *mask2*, are not equal to the previous voxel coordinates' most significant bits (also filtered by *mask2*), the algorithm goes up one level (pop command), and shifts both mask variables to the left. When both most significant bits become equal, the common parent is found and the next part of the algorithm will be executed to descend the octree using the variable *mask1*. The *mask2* variable is shifted left to be able to filter the most significant bits of the coordinates for the octree level immediately upper to the current level. The variable *mask1* is shifted left to be able to filter the correct coordinate bit corresponding to the resulting octree level when the the common parent is finally found. The variable *mask1* is then used to descend the octree.

The next part of the algorithm descends the octree from the common parent *cell*, creating new *cells* when it does not yet exist (when $*pcell = 0$).

4. Visualizing the Octree Using Smart Points

Voxels are stored in an octree, thus, allowing quite huge discrete spaces without a very high memory consumption. Normal vectors are calculated during the voxelization by evaluating the gradient in the middle of the voxel and then normalizing it. A voxel, located at the leaf octree level, is just a pointer to a structure containing the three normal vector components, colour and other information. Higher octree level nodes contain only octree children pointers, when they exist, or zero otherwise. All the voxels are considered as points and rendered using GL or OpenGL from Silicon Graphics Incorporation.

This visualization method can allow close-ups of the surface using levels of details. Levels of details are quite natural to hierarchical voxel models, the models used in this article, because the transition between the original and refined model is indistinguishable.

A major advantage of our method is that no special hardware is necessary to use voxels. In addition, it allows the mixing of polygonal models (for representing polygonal objects) with voxels at graphics engine level, thus, eliminating the need to convert polygons to voxels while profiting from hardware rendering for polygons.

The algorithm describing the visualization technique is given in figure 4.1. The variables *cell* and *root* have initially the address of the root of the octree. Variable i is an index varying from 0 to 7 used to access the current octree element into an eight elements cell. These eight elements identify eight equal sided neighbour cubes, defining a recursive subdivision of a single cube. Each of these elements contains a pointer to a new cell, when this cell contains any part of the surface, or a null pointer otherwise. The recursion is controlled by a stack denoted by the instructions push (to introduce a value in the stack) and a pop operator (to extract a value from the stack). The variable i is assigned a zero value denoting a left to right tree traversal. Both, *cell* and i are pushed in the stack to start the recursive traversal. The recursion is implemented by the do-while loop as shown in the algorithm. The first part inside the loop ascends the tree if i reaches an index greater than 7. Since i is zero in the beginning of the algorithm, the control passes immediately to the second part which descends the tree. This part is a while loop which takes place while $i \leq 7$, indicating that this part also advances to all the elements of the current cell from left to right. The voxel coordinates X, Y and Z are built, bit by bit, from the i values. Notice that the previous coordinates bits are saved by shifting them to the left at each new interaction.

If the current cell is a *leaf* node, then X, Y and Z contain the complete coordinates of the voxel to be displayed and the current element ($cell[i]$) contains a pointer to the normal vector of the voxel. These informations are sent to the graphics card using GL point primitives to display the point with the normal vector. In practice, these informations are first stored in a list and when the list is full all the points are displayed at once to increase efficiency. These details are omitted in the algorithm. Notice that after displaying the

```
cell = root = octree root cell address;
i=0;
push(cell);
push(i);
do {
  /* ascend the octree until i<8 */
  while ((i>7) and (cell≠root)) {
     pop(i);                 /* Ascend one*/
     pop(cell);              /* octree level.*/
     X=X>>1;  Y=Y>>1;  Z=Z>>1;
  }
  while  (i≤7) { /* Descend or move right */
  aux=cell[i];
  X=(X<<1) or (i and 1);         /* Calculate */
  Y=(Y<<1) or ((i>>1) and 1);  /* the voxel */
  Z=(Z<<1) or ((i>>2) and 1);  /* coord.    */
  if (aux=Leaf Node) {           /* Leaf?     */
     Display voxel (X,Y,Z) as a point with the
     normal vector pointed by "aux";
     i=i+1;                      /* Go right */
     X=X>>1;  Y=Y>>1;  Z=Z>>1;
  }
  else {                         /* Not Leaf!*/
     if (aux≠0) {                /* Empty?   */
        push(cell);             /* Descend1 */
        push(i+1);             /* level    */
        cell=aux;
        i=0;
     }
     else {                      /* Empty !  */
        i=i+1;                   /* Go right */
        X=X>>1;  Y=Y>>1;  Z=Z>>1;
     }
  }
 }
} while (cell≠root)
```

Fig. 4.1. Visualization algorithm.

178   N. Stolte

voxel, i is incremented to advance to the next element to the right of the current element. Also notice that the coordinate variables must be shifted one bit to the right.

If the cell does not correspond to a *leaf* node, and if the current element ($cell[i]$) is zero, the element does not exist, therefore the algorithm advances to the next element (by incrementing i) and shifts the coordinates one bit to the right. However, if the current element is not zero, the address of *cell* and the next element index ($i+1$) are saved in the stack, and the algorithm descends the tree by attributing to *cell* the address contained in the current element ($cell[i]$) and making i equal to zero (to restart from the extreme left side again in the new cell).

Once i reaches the value 8, which happens when all the elements of a cell were visited, the control is passed again to the main loop that continues if $cell \neq root$. This time $i > 7$, and the first while loop takes the control. This loop extracts from the stack: (1) the indexes i of the current elements and (2) the cell addresses corresponding to all those cells that were already completely visited. At each interaction this loop also shifts the coordinates one bit to the right. Notice that the loop either stops when a cell not yet completely visited is found (denoted by i values less or equal to 7) or when the root cell is found. If the root cell is found and i is greater than 7, all cells in the tree were visited and the algorithm finishes. Currently this method allows interactive visualization for easy surface inspection.

5. Voxelization

To create the octree, the objects need first to be converted to voxels. This process is called voxelization. There are voxelization algorithms for various kinds of objects: polygons, parametric surfaces and implicit surfaces.

These objects can be integrated with remotely sensed depth maps in the same voxel volume. The depth maps also have to be converted to voxels. This conversion, however, is straightforward.

Depth maps are normally composed by two files. The two files represent a matrix of pixels of a satellite image. One of the files is the image itself that is normally used as a texture to be applied in the model. The other file is another matrix of pixels of the same resolution as the previous file. However, instead of representing the colour/intensity of the pixel it contains the height this pixel has in relationship to the ground. In this way, each pixel is really a voxel, where the X and Y coordinates are given by the image's lines and columns and the Z coordinate is given by the pixel indexed by X and Y.

In this sense, there is no conversion, since depth maps are already voxels, just represented in a different way. However, difficulties exist. The Z coordinates of the depth map can change very quickly from one pixel to another. If just only one voxel would be generated for each map's pixel, holes could show up in the voxel model. For this reason, for each pixel in the map, all the eight neighbour pixels must be analysed. If the neighbours' heights are never more than one unit different than the pixel's height, then only one voxel is generated for that pixel. Otherwise, new voxels have to be piled under and/or over the original pixel to fill the height gaps.

The normal vectors have to be calculated by central difference as explained in section 2. The voxel can also store the texture image given by the first depth map file.

6. Conclusion

This article has shown several techniques to allow visualization of remotely sensed depth maps using voxels. The visualization technique can also be used for visualizing other kinds of objects. The objects have to be first converted to voxels to be visualized. It is shown that the conversion from depth maps to voxels is straightforward. The texture file from the depth map can be directly stored into the voxels. Therefore, no special care is necessary on how the texture mapping will be handled. The approach presented here is highly appropriate for visualizing and manipulating depth maps, since they are basically surfaces. The underground information can be stored in the octree using no extra memory.

An intrinsic advantage of this approach is that objects can be manipulated or edited as in a drawing program on a 2D screen. The volume is a 3D screen that behaves exactly as the 2D counterpart.

Hence, voxels are a much more natural way to represent depth maps. At the same time high quality visualization is achieved at interactive speeds. LCD glasses can be easily integrated into the system for virtual reality applications.

References

1. D. Cohen and A. Kaufman, "3D scan-conversion algorithms for linear and quadratic objects", *Volume Visualization*, pp. 280–301, 1990.
2. T. Duff, "Interval arithmetic and recursive subdivision for implicit functions and constructive solid geometry", *Computer Graphics*, vol. 26, pp. 131–138, July 1992.
3. R. Endl and M. Sommer, "Classification of ray-generators in uniform subdivisions and octrees for ray tracing", *Computer Graphics* forum, vol. 13, pp. 3–19, March 1994.
4. A. Fujimoto, T. Tanaka, and K. Iwata, "ARTS: Accelerated ray tracing system", *IEEE Computer Graphics and Applications*, vol. 6, no. 4, pp. 16–26, 1986.
5. A. S. Glassner, "Space subdivision for fast ray tracing", *IEEE Computer Graphics and Applications*, vol. 10, no. 4, pp. 15–22, 1984.

180   N. Stolte

6. D. Jevans and B. Wyvill, "Adaptive voxel subdivision for ray tracing", in *Proceedings of Graphics Interface '89*, (Toronto, Ontario), pp. 164–172, Canadian Information Processing Society, June 1989.
7. D. Kalra and A. Barr, "Guaranteed ray intersections with implicit surfaces", *Computer Graphics*, vol. 23, pp. 297–306, July 1989.
8. A. Kaufman, "An algorithm for 3D scan-conversion of polygons", in *Eurographics '87*, (Amsterdam), pp. 197–208, North Holland, August 1987.
9. A. Kaufman, "Efficient algorithms for 3D scan-conversion of parametric curves, surfaces, and volumes", *Computer Graphics*, vol. 21, pp. 171–179, July 1987.
10. A. Kaufman, D. Cohen, and R. Yagel, "Volume graphics", *IEEE Computer*, vol. 26, pp. 51–64, July 1993.
11. J. Snyder and A. Barr, "Ray Tracing complex models containing surface tesselations", *Computer Graphics*, vol. 21, no. 4, pp. 119–128, 1987.
12. N. Stolte and Re. Caubet, "Discrete ray-tracing of huge voxel spaces", in: *Eurographics 95*, (Maastricht), pp. 383–394, Blackwell, August 1995.
13. N. Stolte and R. Caubet, "Comparison between different rasterization methods for implicit surfaces", in: *Visualization and Modelling*, (R. Earnshaw, J. A. Vince and H. Jones, eds.), ch. 10, pp. 191–201, Academic Press, April 1997.
14. G. Taubin, "Rasterizing algebraic curves and surfaces", *IEEE Computer Graphics and Applications*, pp. 14–23, March 1994.
15. R. Yagel, D. Cohen, and A. Kaufman, "Discrete ray tracing", *IEEE Computer Graphics and Applications*, vol. 12, no. 5, pp. 19–28, 1992.

Three Dimensional Surface Registration of Stereo Images and Models from MR Images

Henning Nielsen[1], Lasse Riis Østergaard[1], and David Le Gall[2]

[1] Department of Medical Informatics and Image Analysis
Aalborg University, Denmark.
e-mail: hn@vision.auc.dk
[2] Ecole National Supérieure de Physique de Strasbourg
Universite Louis Pasteur Strasbourg, France.

Summary. In image guided brain surgery it is common practice to attach a stereo-tactic frame to the patients skull both during MR scanning and the operation. To avoid the use of this frame we have initiated studies towards the design and implementation of a system to integrate surface information from stereo images with MR images. The goal of the study is to investigate if it is possible to map a discrete set of surface points from the head, obtained from two cameras and structured light, with a surface model made from MR images of the same patient. The requirement being an accuracy of less than 2 mm. The cameras are calibrated using Tsai's calibration method. The projection of the grid on the face gives app. 100 crossing points, which can be seen from both cameras. The registration is done using a distance map in three dimensions and an accuracy of 1.6 mm is obtained.

1. Introduction

The development of computer equipment for use in the health care area has for the last few decades been increasing. One of the huge application domains is *Computer Assisted Surgery* (CAS). The notion CAS covers the use of computer equipment to aid the surgeon before, during, and/or after the surgical intervention. Often a large amount of digital images of the patient is used in CAS. By analysing the images it is possible to e.g. localise a tumour in the brain and plan the operation by determining the path to the tumour. This subarea within CAS is called *Image Guided Surgery* (IGS).

A typical situation prior to a neurosurgical intervention is, that a number of images from different scanner systems, e.g. Magnetic Resonance (MR) and Computed Tomography (CT), is presented to the surgeon. These images show the different brain tissue structures, hence the diseased tissue can be localised. However, this localisation is based on the surgeons ability to mentally relate the image information with the real patient. Modern computer systems enable the surgeon to mark points of interest such as the centre of the diseased tissue (target) and the entry point in digital images using pre-operative planning software. The plan can then be transferred to a device which will help the surgeon navigate to the target in the real patient.

In order to be able to relate the marked points in the digital images with the real patient a stereo-tactic frame is attached to the skull of the patient, thereby obtaining a fixed reference coordinate system. Unfortunately

182   H. Nielsen *et al.*

the stereo-tactic frame is inconvenient for the patient and it is unpractical for the surgeon when doing the surgical intervention. Therefore, there is a high need for a frameless system which will enable the surgeon to relate image information with the real patient. Relating different image information is often referred to as *registration* and has to do with spatial alignment of image information. Much research has been done in the field of medical image registration as new modalities are available, providing the surgeon with the possibility of e.g. combining functional information from PET (Positron Emission Tomography) with anatomical information from MR.

In [1] a survey of recent publications concerning medical image registration techniques is presented. The paper states that of the more promising methods surface-based methods appear most frequently in the literature. It is relatively easy to obtain a surface from the patient, either using laser scanning, probes, 2D imagery etc. and the nature of the surface-based methods makes them computational tractable.

In [2] an automatic technique for registering segmented clinical data with any view of the patient on the operating table is proposed. A laser range scanner is used to collect 3D data of the patient's skin as positioned on the operating table. The 3D data is then registered to a segmentation of the skin based on a sequence of MR images. This enables a mix of live video of the patient with a segmented 3D model of the brain. Using 481 laser sample points a registration accuracy of 1.6 mm is obtained.

We propose a registration method using only a few 3D surface points of the patients skin obtained using stereo vision and a segmentation of the patients skin based on a sequence of MR images. The stereo images are acquired using standard off-the-shelf CCD cameras with standard lenses. By projecting structured light onto the face of the patient and collecting 3D points based on a camera calibration it is expected that a registration to the MR skin segmentation can be obtained with an accuracy less than 2 mm.

The paper is organised as follows. In section 2 the extraction of 3D points from the segmented MR sequence and the stereo vision setup is described along with the registration algorithm. Results of the registration process are presented in section 3, in terms of the registration accuracy and a discussion is given in section 4.

2. Methods

2.1 Surface Model from MR Sequence

MR images are slice images with a high spatial resolution within the frame (x, y), but the spatial resolution in the z direction perpendicular to the image is limited by the distance between the slices. In our case the resolution in the x and y direction is 1 mm and in the z direction it is 3 mm due to the fact that smaller slice thickness gives less signal to noise ratio. However, the gray scale contrast between soft tissue and air is fairly big, so segmentation between air and soft tissue can easily be performed by global thresholding. After thresholding a few small blobs in the background are classified as being soft tissue. These blobs are removed by two dimensional morphological opening followed by closing with a square structuring element of size 3 × 3. After this operation there still exist some areas within the head being misclassified as air. These voxels belong to bone tissue, which in all MR scannings has a very low gray level value. Labelling the black part of the whole data set, and in each frame only selecting the biggest object, produces a data set where all voxels inside the skin is one big connected component in 3D. From this new dataset voxels belonging to the persons surface are found as the difference between the dataset and an erosion of the same dataset. Figure 2.1 shows one frame from the MR Image sequence and the contour made from this frame. Only the upper 80 lines of the contour image are used, as this is the part of the face that can be seen from the two video cameras. In figure 2.2 all 65 contours are put together to form the tree dimensional surface of the head.

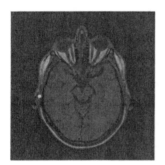 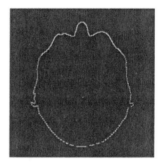

Fig. 2.1. Left) MR image. Right) Contour of MR image.

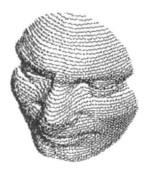

Fig. 2.2. 3D surface points of the face.

2.2 Three Dimensional Points from Video Images

3D points from the face of the patient are found using a stereo setup with two video cameras and a light source with a grid pattern. From both images the crossing points on the face are found, and from corresponding points in the left and right image a set of 3D points are calculated using triangulation. In order to be able to use triangulation we need to know the position, orientation, focal length and the radial distortion of the two cameras. This is achieved by first taking two images of a calibration object, one with each camera. The camera calibration is made by using a software package from Tsai [7] (available on the web [8]). One of the advantages by using this software is that the camera parameters are estimated from many image points and thereby reducing the influence of noise in the position of the individual calibration points. For the camera calibration we use a calibrated object with six plane surfaces of 68 calibrated points and a binary code pattern on top of each side, which indicates to the software, which side it is (see figure 2.3).

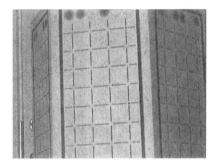

Fig. 2.3. Calibration object as seen from the left camera.

The calibration object is made at our laboratory and further details can be found in [9]. An image of a person illuminated with a grid pattern, can be seen in figure 2.4.

There are several ways of finding the relevant crossing points. One way is by looking for cross points using a cross detector based on template matching. Another one is based on first finding the horizontal and the vertical lines. The crossing points are then found by a simple AND operation between the two sets of lines. We have tried both methods and found that the line based method is the fastest and best method. Only points on the skin (not in hair and beard) that can be seen from both cameras are selected for use in 3D triangulation. In this project we found 90 useful points on the skin.

2.3 Three Dimensional Distance Map from MR Sequence

The final matching between the MR surface and the points on the face, found from triangulation, requires calculation of the minimal distances between ev-

Fig. 2.4. (a) Image seen from the left camera. Person illuminated with structured light. (b) Horizontal lines. (c) Vertical lines. (d) Crossing points used in triangulation.

ery crossing point on the face and all surface points from MR. The complexity of this operation is very big, approximately one million distance calculations for every iteration of the matching algorithm, so we need a faster algorithm for finding the minimum three dimensional Euclidean distance from a point to a surface. From the image processing literature it is well known, that the minimum Euclidean distance from an object can be found by four scans through a two dimensional image [3, 4], and as early as 1984, Gunilla Borgefors described how distance transforms can be made in any dimensions [5]. In 1996 Claus Gramkow from IMM[1] [6] made a program for the calculation of the minimum Euclidean distance in three dimensions. This program was used for the calculation of the minimum Euclidean distance to the surface of the head in all voxels within a box of size equal to the size of the face ± 20 mm in all three directions.

2.4 Registration

The two data sets in the registration are the surface points from MR and the points found from the grid-pattern on the face. The first set we denote P_{MR} and the second set P_v. Both sets can be seen as a rigid body, and therefore only translation and rotation are allowed to register P_v to P_{MR}. Neither scaling nor warping of one data set to the other can be used. The data set P_{MR} represent the surface points from the face, and only the frontal part of the head is used. A three dimensional matrix D consists of all voxels within a box of size equal to the face ± 2 cm in all directions, and $D(x, y, z)$ are the minimal Euclidean distances from point (x, y, z) to the surface P_{MR}. The number of points in P_{MR} is typically 5000. From the set of points defined by the grid on the face only points seen by both cameras are used. Points in the beard or the hair are not used since they can not match points on the surface of the MR data. The data set P_v typically consists of 50 to 100 points depending on the spacing between the grid lines. The registration is correct

---

[1] Department of Mathematical Modelling, University of Denmark

186   H. Nielsen *et al.*

when all points from P_v coincide with points from P_{MR}. This is not possible as P_{MR} are integers with steps of $1 \times 1 \times 3$ mm in the x, y and z direction, and P_v are floating point 3D points. So the registration is a matter of rotating and translating the rigid body defined by P_v until best match with P_{MR} is achieved. As a measure we use the sum of all the squared minimum distances from $P_v(i)$ to P_{MR}. However, these minimum distances from all points to P_{MR} are defined in the matrix D, so the error measure is found by, for every translation or rotation, calculating the sum of the squared values from D addressed by P_v,

$$\epsilon = \frac{1}{N} \sum_i D(P_v(i))^2 \tag{2.1}$$

where, N is the total number of points in P_v. For normalisation purposes these error measures are divided by the number of points in P_v.

$\mathbf{M}$ is a 4×4 matrix defined by

$$\mathbf{M} = \left(\begin{array}{c|c} r & t \\ \hline 0 \ 0 \ 0 & 1 \end{array} \right) \tag{2.2}$$

where r is the rotation part and t is the translation.

The algorithm for finding the matrix M that transforms P_v to P_{MR} is split into two parts.

A. Perform an initial translation and rotation of P_v so that all points from P_v address points defined in D. The translation part of this can be found as the difference between the mean x, y and z values from P_v and P_{MR}. The rotation part can be found by rotating P_v around its mean point around all three axes until all elements from P_v are within the definition range of D. This rotation can be tested in steps of 5 degrees as this is small enough to assure that all points are within $P_{MR} \pm 2$ cm, which is the definition range of D.

B. After having initialised P_v to be within D, the final matching is performed by iteratively testing translations of ± 1 pixel in all three directions and for every iteration updating $\mathbf{M}$ with the step giving the minimum error, until it converges. The maximum number of steps being < 20 as the range is 20 mm. After fine-tuning the translation, the rotation part of $\mathbf{M}$ is iterated in steps of 0.5 degrees around all three axes until minimum error is found. This iteration will converge within 10 steps ($5° / 0.5°$).

3. Results

The objective of this investigation was to see, if it was possible to find a rigid body translation and rotation between a surface from MR images and the same surface obtained from stereo imaging. The requirement being that the registration should match within 2 mm. In order to evaluate the result we have to look at the distribution of the 90 distances between the translated and rotated points $P_v = M * P_v$. The mean distance is 1.6 mm, the maximum distance is 4.5 mm and the distribution of distances are as shown in figure 3.1. 61% of all points are within ± 1 pixel in all directions, and 97% are within ± 2 pixels, as a diagonal pixel step in 3D is $\sqrt{3}$ mm. The first method where we calculated the minimum distance from P_v to P_{MR} took 10 hours for one iteration, and therefore this method was not investigated any further. The second method, which we actually used, was based on pre-calculation of a 3D-matrix containing all the minimum Euclidean distances from the face. This calculation took 2 minutes, the only disadvantage being that the size of this matrix is 780.000 floating point numbers. The iterative matching between the two data sets was done using the Matlab software tool and it takes less than 2 sec. for a total of 15 iterations. All calculations are made on a standard O2 workstation from SGI.

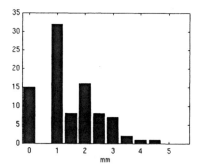

Fig. 3.1. Error distribution.

4. Discussion

The purpose of this study was to see if it was possible to use structured light and standard video cameras instead of a stereo-tactic frame to relate MR Images to the patient on the operation table. The goal was that the registration should be within ± 2 mm. We used a method based on a pre-calculated distance map in three dimensions relative to the face of the patient. It took 2 minutes to make the distance map, and this can be done prior to the operation. The matching between the two data sets, from MR Images and video triangulation, took less than 2 sec. The mean Euclidean distance from all 90 points to the surface of the patient was found to be 1.6 mm. So the goal was fulfilled. When selecting the field of view from the two cameras, it is important that the surface contains some "hills and valleys". e.g. the nose

188   H. Nielsen *et al.*

and the eyes. Otherwise the matching may give several solutions. It should be possible to use this method on other types of data e.g. different terrain models of a landscape that has to be aligned.

References

1. J. B. Antoine Maintz and M. A. Viergever, "A survey of medical image registration", *Medical Image Analysis*, vol. 2, no. 1, pp. 1–36, 1998.
2. W. E. L. Grimson, G. J. Ettinger, S. J. White, T. Lozano-Perez, W. M. Wells III, and R. Kikinis, "An automatic registration method for frameless stereotaxy, image guided surgery, and enhanced reality visualization", *IEEE Transactions on Medical Imaging*, vol. 15, no. 2, pp. 129–140, 1996.
3. E. Piper and E. Granum, "Computing distance transforms in convex and nonconvex domains", *Pattern Recognition*, vol. 20, pp. 599–615, 1987.
4. U. P. V. Skands, *Template matching by mathematical morphology*, IMM-DTU, M.Sc. Thesis, 1992.
5. G. Borgefors, "Distance transformations in arbitrary dimensions", *Computer Vision, Graphics, and Image Processing*, vol. 27, pp. 321–345, 1984.
6. C. Gramkow, *Registration of 2D and 3D medical images*, IMM-DTU, M.Sc. Thesis, 1996.
7. R. Y. Tsai, "A Versatile Camera Calibration Technique for High-Accuracy 3D machine vision metrology using off-the-shelf TV cameras and lenses", *IEEE Journal of Robotics and Automation*, vol. RA-3, no. 4, 1987.
8. http://www.cs.cmu.edu/~rgw/TsaiCode.html
9. J. Bjørnstrup, *Automatic camera calibration using a passive calibration object*, LIA-AAU, Technical Report, 1998.

Exploring Multi-Dimensional Remote Sensing Data with a Virtual Reality System

Walter Di Carlo

Space Applications Institute, Joint Research Centre,
European Commission, I-21020, Ispra, Varese, Italy.
e-mail: walter.dicarlo@jrc.it

Summary. A scientific visualization tool, based on a virtual reality system, has been created to explore remotely sensed data. The main aim is to find possible relationships between results of different spectral unmixing techniques in order to improve image classifications. In particular, data from fuzzy classifiers and linear spectral unmixing techniques, has been considered. The tool offers the possibility to use different representations for high dimensional data due to the fact that none of the existing representations is always effective to visualise multi-dimensional data. Furthermore, the possibility to visually link different representations of the same data, by using a technique called data brushing, will help the analyst in to get more insight into the structures present in the high dimensional data space and, to find relationships between the data spaces. The exploratory tool has been based on a virtual reality system in order to create new human-computer interaction paradigms. Furthermore, it has been used to assess the usability of immersive computer interfaces, called virtual environments, in data exploration. The main advantages in using this kind of interface has been an increased interface space and a more natural human-computer interaction which means more information content requiring less time to become familiar with the tool.

1. Introduction

In general, the visualisation process may facilitate the understanding of complex data sets, such as remotely sensed data, by preserving as much essential information as possible from the original data set. However, in many cases, a single visualization technique is not sufficient to extract all properties of the data or cannot be used to explore different types of data. For example, as described later in the paper, parallel coordinate plots, which are a visualisation tool for multi-dimensional data sets, cannot be used to extract trends.

In this context, a graphical tool has been developed to assess the usefulness of visualisation techniques for exploration of remotely sensed data. Selections of existing techniques have been tested both independently and in combination with each other. The objective of this study is to gain insight about techniques used in remote sensing to extract sub-pixel information, such as those based on fuzzy classifiers [7] and linear spectral unmixing [1, 3], in order to improve image classification. The main idea of the above techniques is that the spectral signatures of remotely sensed images are mixtures of different elements. This means that they cannot be associated with just one class, as done by so-called hard classification techniques. In particular,

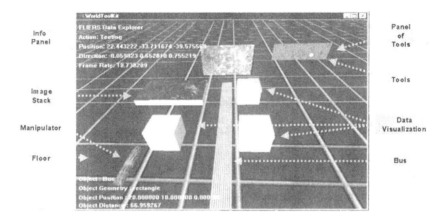

Fig. 1.1. Virtual Environment Overview

fuzzy classification considers each pixel associated with different classes with different strengths using fuzzy sets. Spectral unmixing, on the other hand, considers each pixel as a combination of some fundamental pure components called "end-members" such as soil, vegetation, shadow etc.

The graphical tool uses an immersive 3-dimensional interface (figure 1.1), called "virtual environment", created with a virtual reality system. The main reasons for adopting this type of interface are the possibility to display more data than in standard interfaces [4] and a more natural human-computer interaction.

2. Visual Data Exploration

In remote sensing, the information is both in spatial and multi-dimensional feature space and in many cases they are "linked" to each other. For example, a multi-band image might be considered as a set of spectral layers (bands) or as a matrix of spectral signature vectors. Hence a pixel in a certain position in a layer is "linked" with the spectral vector having the same position in the corresponding matrix. In other cases the information is organised as a hierarchical structure, such as classified data sets, where classes could be organised as a tree of relationships. As a consequence, a set of graphic techniques has been chosen in order to display spatial and multi-dimensional data sets such as:

- Spectral bands. The original data highly correlated in spectral and spatial space [9].
- Earth features fuzzy sets. These data sets are the results of fuzzy classification techniques applied to a set of spectral bands. In this case, data are represented by membership images with pixel values in the range [0,1].

- Texture features information. This data set consist of images (texture feature images) obtained by applying different texture analysis methods [12].
- Spectral unmixing abundance data. These data consists of images derived by spectral unmixing techniques. Each image, called an abundance image, represents the contribution of each end-member [1, 3].
- Testing data. Labelled or reference data used to verify the results of the different classification techniques. In this case, tables are used to associate with each multi-spectral data a label indicating the land cover type [9].

The main goal of a tool for data exploration is to gain more insight about the structures present in the data, but each single existing technique used to visualise multi-dimensional data cannot be used for different aims or different data sets [13]. Furthermore, in exploring multi-dimensional data one must also consider that [5]:

- In many cases, the dimension of the structures present in high dimensional data sets is independent from the size of the data space containing the data.
- As the dimensionality of the data increases, the geometry characteristics exhibit surprising behaviour such as the fact that most of the data are near the corners of the hyper cube or that most of the data are in the outside shell of any hyper sphere.
- Most of the data space is empty Hence, a technique called data brushing has been adopted in order to have many different views of the same data by using different visualisation techniques "visually linked" to each other. This means that the user can select some data points in a graph and see the same data points highlighted in all linked graphs. This could be used to explore relationships between spatial and feature spaces or between different feature spaces.

The tool for the visual exploration of remotely sensed data uses the following graphs:

Image Stack. A set of images coming, for example, from bands of a multi-spectral remotely sensed image, fuzzy classification etc., is displayed as a stack of layers which one can use to view/select spatial information such as polygons and/or pixel regions.

Hyper-Corner Plot. This visual tool displays data using as coordinates their distance from selected corners of the data space's hyper-cube [8]. This technique helps to see data from the corners of the data hyper-cube space where they tend to be in high dimensional data space.

Parallel Coordinate Plot. Here, the points in the high dimensional space are displayed as a broken line, snaking along the parallel axes (figure 2.1). In this case, the graph depends on the order of the axes and it is possible to detect enclosed surfaces and intersecting trajectories present in the high dimensional space, although the correspondence may not be obvious at first. Relationships

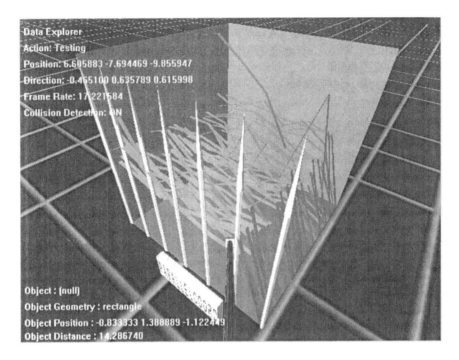

Fig. 2.1. Parallel Coordinates plot.

between adjacent dimensions manifest themselves as line segments with similar orientation. Adjacent dimensions with negative correlation create an "X" pattern between them. Furthermore it is possible to do simple point pair wise comparison but it is not always possible to locate data outliers. With this technique, it is not possible to see trends in data. Note that a point in the Cartesian space is represented as a snaking line while a Cartesian line is dually represented by an intersection point of many snaking lines. Furthermore, parallel lines are represented in the parallel coordinate plot as intersection points that lie on the same vertical line. Note that, the 3-dimensional version of the parallel coordinate graph uses the third axis to order data points and to reduce data overplotting.

2D/3D Glyph Plot. A glyph is a graphical object whose attributes – position, size, shape, colour, orientation, texture etc. – are bound to the data. In many cases, glyphs are used together with other techniques to increase the information content of graphs [11]. In particular, they are used to look for similar shapes and simple transitions between shapes. Furthermore it is also fairly easy to spot anomalies.

3. Visual Data Analysis

The main goal of scientific visualisation is to aid the understanding of complex data sets while preserving as much essential information as possible. Typically, a tool for data exploration and analysis is used to become familiar with the data, discover trends and relationships, and perform comparisons. The virtual reality tool described in section 2 will be used mainly for:

- Selection of best band combination to reduce data quantity and increase data quality using a multi-dimensionality reduction technique, such as "exploratory projection pursuit".
- Investigation of relationships between multi-dimensional feature and spatial spaces to find possible data properties. The exploratory projection pursuit technique might be used to become familiar with the feature space and to find interesting structures. On the other hand, the hyper-corner plot might be used to get an idea of feature space distribution. The parallel coordinate plot can be used to see the spectral values of the data. In the same way, glyphs might be used to increase the information content of each graph. On the other hand, the image stack might be used to display the spatial information.
- Identification of outliers to increase data quality and therefore image analysis and classification accuracy.
- Visual end-members selection and verification. By connecting end-members with lines it is possible to evaluate their accuracy by evaluating the number of data points that lie outside the polygon created by the projection of their connections. The parallel coordinate plot of the abundance of end-members might be used to select the end-members and to evaluate their accuracy.
- Investigation of possible relationships between fuzzy classification and spectral unmixing.

4. Immersive Virtual Environment

An immersive virtual environment is an application created with a virtual reality system to display and interact with a 3-dimensional representation of data [10]. In general, virtual reality systems consist of a set of hardware and software units necessary to interact with the virtual environment. The most important are:

- Head mounted displays (HMD). A small monitor for each eye is used to generate a stereoscopic representation of the virtual environment and to give the feeling of immersion inside that environment.
- A digital glove. This is used to interact with the elements present in the 3-dimensional interface.

194 W. Di Carlo

- A set of six-degree-of-freedom sensors. They are used to track the position of the user in order to update its representation into the virtual environment.
- A 3-dimensional software graphics library.

The exploratory tool has been created with a virtual reality system to reduce the overplotting problem. For example, the 2-dimensional parallel coordinate plot, is difficult to understand when too many data points have been displayed. The tool also increases the amount of data available to the user in order to gain more insight from the data. The main advantages of using immersive virtual environment over the standard 2-dimensional interface are:

- To display more data. The sensor mounted on the HMD permits the virtual environment to be easily rotated, increasing the space available to display data. Furthermore, the stereoscopic view reduces the overplotting problem (too many data points) by spreading the data in the third dimension [4].
- To reduce distractions. The head-mounted displays prevents the user from seeing objects outside the virtual environment, reducing the noise of the surrounding real environment.
- To have interaction techniques that are more natural to use. For example, a virtual pen can be used to select data while a virtual video camera might be used to explore the virtual environment.
- To increase the interface feedback using 3-dimensional sound [6].

The interaction with the virtual environment is done mainly by using three type of components:

Virtual tools. A set of virtual tools can be used to interact with the virtual environment (to fly, to move objects, etc.) and the data (select, annotate, etc). The user can use a tool by selecting it from the provided virtual panel of tools.

Data graphs. These are based on the visualization techniques to display spatial and multi-dimensional data sets. In figure 1.1 the graphs are represented by boxes.

Data bus. This is an object (see figure 1.1) used to build a simple data network. Data sources such as multi-band images and data consumers (data visualisation technique) can be attached to the bus. The user might utilise it to connect a data source with different data consumers in order to have different views of the same data and with the possibility to use the data brushing technique.

5. Discussion

A tool for the visual exploration of remote sensing data has been created. Spatial and multi-dimensional data spaces have been taken into consideration. In particular, a technique called data brushing has been adopted which allows the user to select data points in a graph and to view the same data points highlighted in all linked graphs. This is an important feature which allows increased exploration and understanding of the relationships between spatial and feature spaces or between different feature spaces.

The tool is based on a virtual reality system in order to increase the content of the visualisation and, at the same time to reduce the complexity of the interface, even if this type of system is not free from problems such as the:

- Nausea that could be generated from the latency in the visualisation of the virtual environment. This is because computers are still not powerful enough to simulate complex scenes in real-time [2].
- Possibility to get lost in the virtual environment when good reference points such as a floor or reference objects are missing.
- Limited space available for movement due to the cables of the helmet and glove worn by the user.

More experiments are necessary, especially with real data sets, in order to evaluate the usefulness and the performance of the exploratory tool with remotely sensed data. In particular, tests will be carried out to:

- Evaluate the scalability of the different visualization techniques with respect to the number of classes present in the data.
- Understand how the tool will be used by different remote sensing users in order to improve its usefulness and usability.
- Improve the tool by adopting advances visualization techniques such as the exploratory projection pursuit.

Acknowledgement. This work has been supported by the EU Framework IV FLIERS project. Many thanks to my colleague Stefania Goffredo in helping me to write this document.

References

1. J. B. Adams and M. O. Smith, "Spectral mixture modeling: A new analysis of rock and soil types at the viking lander 1 site", *Journal of Geophysical Research*, vol. 91, pp. 8098–8112, 1986.
2. B. Azar, "Psychologists Dive Into Virtual Reality", *APA Monitor Online, The Newspaper of the American Psychological Association*, March 1996. (http://www.apa.org)
3. C. Coll, J. Hill, C. Hoff, S. de Jong, W. Mehl, J. F. W. Negendank, H. Rienzebos, E. Rubio, S. Sommer, J. Teixeira Filho, E. Valor, B. Lacaze, and V. Caselles, *Alternative solutions for mapping soil and vegetation conditions through remote sensing: The spectral mixing paradigm*, chapter 3.3, pp. 69–86. EUR 16448 EN, European Commission, 1996.

196   W. Di Carlo

4. G. Franck and C. Ware, "Viewing a graph in a virtual reality display is three times as good as a 2d diagram", in *IEEE Symposium on Visual Languages*, pp. 182–183, October 1994.
5. M. A. Carreira-Perpinan, "A review of dimension reduction techniques", Technical Report CS-96-09, Department of Computer Science, University of Sheffield, 1997.
6. M. Cohen, S. Aoki, N. Koizumi, "Augmented Audio Reality: Telepresence/VR Hybrid Acoustic Environments", *Proceedings of the second IEEE International Workshop on Robot and Human Communication*, pp. 361–364, 1993.
7. A. Rampini E. Binaghi and P. A. Brivio, eds., *Soft computing in remote sensing data analysis*,Singapore: World Scientific, 1995.
8. D. W. Levandowski H. Cetin and T. A. Warner, "Data classification, visualization, and enhancement using n-dimensional probability density functions (npdf): Aviris, TIMS, TM, and geophysical applications", *Photogrammetric Engineering and Remote Sensing*, vol. 59, pp. 1755–1764, December 1993.
9. T. M. Lillesand and R. W. Kiefer, *Remote sensing and image interpretation*, New York: John Wiley & Sons, Inc., third ed., 1994.
10. M. R. Mine, "Virtual environment interaction techniques", Technical Report TR95-018, University of North Carolina, Chapel Hill, Nc 27599-3175, May 1995.
11. G. G. Grinstein and R. M. Pickett, "Iconografhic displays for visualizing multidimensional data", in *Proceedings, IEEE Conference on Systems, Man and Cybernetics*, pp. 514–519, 1988.
12. J. A. Richards, *Remote sensing digital image analysis*, Berlin: Springer-Verlag, second, revised and enlarged edition ed., 1994.
13. C. Schmid, "Comparative multivariate visualization across conceptually different graphic displays", in *Proceedings of the Seventh International Working Conference on Scientific and Statistical Database Management*, 28–30 September 1994, Charlottesville, VA, IEEE Computer Society Press, pp. 42–51, 1994.

Part IV

Image Interpretation and Classification

,

Information Mining in Remote Sensing Image Archives

Mihai Datcu[1], Klaus Seidel[2], and Gottfried Schwarz[1]

[1] German Aerospace Centre – DLR, German Remote Sensing Data Center – DFD,
Oberpfaffenhofen, D-82234 Wessling, Germany.
e-mail: mihai.datcu@dlr.de
[2] Swiss Federal Institute of Technology – ETH Zürich,
Communication Technology Laboratory,
Gloriastr. 35, CH-8092 Zürich, Switzerland.
e-mail: seidel@vision.ee.ethz.ch

Summary. Conventional satellite ground segments comprise mission operations and data acquisition systems, data ingestion interfaces, processing capabilities, a catalogue of available data, a data archive, and interfaces for queries and data dissemination. The transfer of data mostly relies on common computer networks or high capacity tapes.

Future ground segments have to be capable of handling multi-mission data delivered by Synthetic Aperture Radar (SAR) and optical sensors and have to provide easy user access to support the selection of specific data sets, fuse data, visualize products and to compress data for transmission via Internet. In particular, the search for data sets has to support individual queries by data content and detailed application area ("data mining") as well as capabilities for automated extraction of relevant features and the application oriented representation of results. In the case of SAR image data, we face an enormous volume of raw data combined with very specific analysis requests posed by the users. In order to reconcile these conflicting aspects we suggest the development of a tool for interactive generation of high level products.

This tool shall extract, evaluate, and visualize the significant information from multidimensional image data of geographically localized areas. It shall rely on advanced image compression methods. Thus, the tool will support the monitoring of the Earth's surface from various perspectives like vegetation, ice and snow, or ocean features.

This exchange of information can be understood as image communication, i.e. the transfer of information via visual data. We propose an advanced remote sensing ground segment architecture designed for easier and interactive decision-making applications and a broader dissemination of remote sensing data.

This article presents the concept developed at the German Aerospace Center, DLR Oberpfaffenhofen in collaboration with the Swiss Federal Institute of Technology, ETH Zurich, for information retrieval from remote sensing (RS) data. The research line has as very pragmatic goals the design of future RS ground segment systems permitting fast distribution and easy accessibility to the data, real and near real-time applications, and to promote the implementation of data distribution systems. However, the scope of the project is much larger, addressing basic problems of image and information representation, with a large potential of application in other fields: medical sciences, multimedia, interactive television, etc.

1. Conventional Remote Sensing Data Ground Segments

Conventional ground segments comprise mission operations and data acquisition systems, data ingestion interfaces, processing capabilities, a catalogue of available data, a data archive, and interfaces for queries and data dissemination. The transfer of data mostly relies on common electronic networks or high capacity tapes. Disregarding the production and handling of different product levels, a simplified architecture of a conventional remote sensing data ground segment from the perspective of information transfer is shown in figure 1.1 [22].

The remote sensing data, i.e. raw data, images or value added products, are stored in the physical archive. At insertion, the data ingestion engine extracts the ancillary information (e.g. sensor mode information, time and geographical location of data acquisition), and also generates quicklook images. A typical remote sensing data file has a size in the range of 100 MBytes (ERS-1 scene), a typical quicklook file comprises 100 KBytes. The quicklook file size reduction is several hundreds of times. The existing algorithms for quicklook generation are a simple resampling of the original data; consequently, the information content of an image is strongly degraded.

In general, the ancillary information is archived using relational data bases in an inventory. This is the only information available for a query in the image archive. The users specify their queries via a query engine. The result of a query process is a set of links (pointers) to quicklooks and the corresponding full resolution images being all similar in the sense expressed by the query (e.g. all ERS-1 images in the area of Bonn, acquired during 1994). The response to a query results in a cluster of images - usually tens up to hundreds

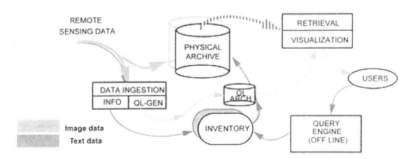

Fig. 1.1. Simplified architecture of a conventional remote sensing data ground segment from the perspective of information transfer. The channels transmitting images are shown in dark gray, the channels transmitting logical data (e.g. ancillary information) are shown in light gray, and the channels subject to commercial restrictions are dashed. The thickness of the arrows is related to the amount of information transferred. INFO stands for the information in the header of remote sensing signal/image files, QL-GEN means quicklook image generation, QL ARCH is the quicklook archive.

of data sets. The user has to browse through the quicklooks and, based on visual information, will select the most useful images.

The last step of the data access procedure is the retrieval of full resolution images from the physical archive. Presently, as a consequence of both commercial policy and conventional technology, the retrieval of full resolution data is an off-line process. Data interpretation and information extraction is done either before archiving, during value adding production (e.g. generation of geocorrected images, biomass or meteorological maps), or at the user site. As an example we present the Intelligent Satellite Data Information System (ISIS) of the German Remote Sensing Data Center – DFD [10]. ISIS, is the link between an elaborate archive system and the users wishing to tap its resources. The users are only involved at the two ends of the long archiving and distribution chain described above: firstly, during a request and, secondly, by receiving information. ISIS internally manages the more tedious links in between which involve translating the user requests into a language understood by the archive and the data management system, and conducting the sometimes extensive interactive dialogue necessary before a meaningful response can be supplied. ISIS provides data catalogue search and retrieval, access to texts and software, and links to external information systems and archives.

In addition, ISIS facilitates network transfer of digital quicklooks for visual inspection, transfer of selected data (full data sets or interactively defined subsets), and electronic order placement. In order to support the entire spectrum of hardware platforms ISIS serves different interfaces and is available via common communication lines. The graphical interface GISIS provides the most comfortable access to the data of the DFD catalogue offering distributed catalogue retrieval from more than 200,000 data entries with about 120,000 digital images (July 1997), supported by a map browser, geographical names, and databases; display of data set footprints on a map; automated network transfer of digital quicklooks for visual inspection; transfer of full resolution data (full data sets or interactively defined subsets) for some data products; guide information about sensors, data products and distribution modalities; online order placement; online placement of user comments (http://isis.dlr.de/).

2. Future Remote Sensing Data Ground Segments

Future ground segments have to be capable of handling multi-mission data delivered by SAR and optical sensors and have to provide easy user access to support the selection of specific data sets, fuse data, visualize products, and compress data for transmission via Internet. New features of such systems are interactive distributed decision systems for data mining, query by image content in large archives, information extraction and interpretation, availability of advanced processing algorithms and computer technology, as well as (near) real time data distribution and dissemination [6].

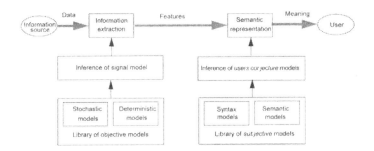

Fig. 2.1. Scalable concept for information retrieval.

The previously discussed concepts led to the design of a new generation of RS ground segment systems resulting in scalable systems for information retrieval:

- query by image content from large image archives,
- data mining in the set of images obtained as a query response, and
- scene understanding from the images found to be relevant for the user's application.

The last system component, scene understanding, is presented elsewhere ([8, 9]).

In figure 2.1 we present a scalable concept for information retrieval from very large volumes of data.

The three levels of the system are designed to accommodate a huge volume of information. They perform a progressive data reduction with simultaneous increase of the image detail extracted.

Due to the huge volume and high complexity of the data to be managed, the newly emerging applications of information extraction systems demand a high level representation of information. The extracted objective signal features are interpreted according to a certain user conjecture. The interpretation process relies on restructuring (using a certain syntax) the signal feature space according to the semantic models of the user. Augmentation of the data with meaning, including the model of user understanding, demands the modelling of the message semantics. The diagram in figure 2.2 presents the simplified architecture of an advanced information retrieval system.

The information source is assumed to be a collection of multidimensional signals, e.g. a multi-mission image archive, or measurements of different physical parameters of a certain process. The information retrieval process is split up in two steps: i) objective information extraction and, ii) semantic representation.

The objective information extraction requires the signal modelling as a realization of a stochastic process [17]. A library of stochastic and deterministic models is used to infer the signal model. The information extraction is

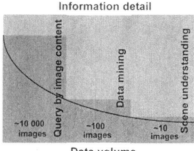

Fig. 2.2. Simplified architecture of an advanced information retrieval system.

the task of incertitude alleviation over the model space and selected model parameters [12]. The resulting objective features, either model types or model parameter vectors, are interpreted according to the user conjecture. The interpretation process relies on restructuring (using a certain syntax) the signal feature space according to the user semantic models.

3. Query by Image Content from Large Remote Sensing Archives

3.1 Inference of Signal Models

The goal of a query by image content is to enable an operator to look into a large volume of high complexity images. An overview of the state of the art can be found in several special issues [1, 2, 3, 4, 5]. The incorporation of content-based query and retrieval techniques into remote sensing image archives will meet current user requirements and, at the same time, lay the foundations for the efficient handling of the huge dataflow that is expected from the next generation of satellites. Establishing the connection from the image signal to the image content is the key step in providing an intelligent query from large remote sensing data bases, allowing the user to obtain model parameters before actually retrieving data.

A crucial problem for content-based image interpretation and manipulation (i.e. archiving, query, retrieval, transmission) is that certain assumptions about the information sources have to be made. These assumptions are made in the form of either stochastic or deterministic models. Thus, an important issue for image interpretation is the process of finding good models for the representation of the information content.

We approach the information extraction of remote sensing images from the Bayesian perspective, i.e. selecting from a pre-defined library of models those models which best explain the image (figure 3.1) [22].

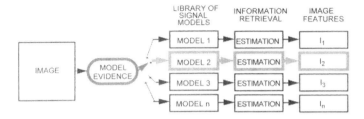

Fig. 3.1. Information retrieval is always a model based approach. We use the models as points of view over our data. Each model explains different features. We define the information retrieval as a two step procedure: model selection, and model parameter estimation. The model type is a meta feature, while the values of the estimated parameters represent the information.

First, the query specification is analyzed and from a library of models the most likely ones to explain the query will be selected. The prior evidence over the model space actually represents in an objective way the user interest in a certain application [20, 21]. We define the information retrieval as a two step procedure: model selection, and model parameter estimation. The model type is a meta feature, while the values of the estimated parameters, the image features, represent the information [6, 9, 18].

Additional information can be extracted using multi-sensor or multi-temporal analysis techniques. Further, structural information extraction as texture features is an important topic. An example is shown in figure 3.2 [18].

Fig. 3.2. A stochastic Gibbs model is used for the extraction of structural information from images. Left: X-SAR scene, Right: RESURS-01 image.

The strategy we adopt is an extension of structural modelling to multiple scales [7, 13, 16, 18]. We define structural meta-features to be used in queries with progressive detail as information at large scales in contrast to information at small scales. The meta-information, e.g. drainage basins, rough mountains, or directional geological alignments will be used to cluster

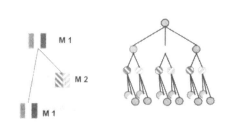 

MULTIRESOLUTION of the MODELS   LABELS of the FEATURES   SIGNAL CLASSES

Fig. 3.3. The utilization of two models (M1 and M2) can explain physically significant structures at multiple scales (pixel, pixel neighbourhood, and scene scale). The same scene is analyzed at several resolutions. M1, at coarse scale understands the image as being composed of large areas (agriculture, mountains, forest and lake). At intermediate scale, M2 recognizes the orientation of the agriculture parcels, and at the finest scale M1 now explains the pixel classes.

the archived images thereby reducing the volume of the next finer query, e.g. type of vegetation, urban area, or snow cover. An example is contained in figure 3.3 [6].

Thus, any structural information extraction at multiple scales has a dual goal:

- increasing the robustness of the two step information extraction process (the scale is defined as a new meta-feature) and
- being the base for a strategy to discard during the query the non-interesting images in a time efficient algorithm.

We experimented with our new content-oriented query method using a small test image archive. We set up a small test database comprising 484 small windows extracted from Landsat-TM images to demonstrate the basic scheme of the proposed archiving method. It was used to verify on an experimental basis how direct, on-line feature matching methods can be used in addition to the indexing approach to query and to retrieve stored data [23]. Our library of models contains two classes of stochastic models. Firstly, multispectral models [11], and secondly, a sub-library of 5 texture models (Gibbs Random Fields) with different degrees of complexity [18, 19]. The query language accepts a vocabulary of 7 elements and, in addition, the order of complexity of the model and a degree of similarity (distance) of the images. A query for fields using a model of order 5 and a relative high degree of similarity (0.2) resulted in finding 4 images from the existing 484 data sets. An on-line demo is available at [23].

The proposed architecture is designed to keep compatibility with current image query and retrieval methods used in remote sensing. However, in addition to the conventional approach based on queries for sensor type, location and time of acquisition, we intend to support searching by image content. For this purpose, we investigated the use of general, robust image features to be used as indexes in a relational database. In contrast to similar approaches, these features will not characterize application specific aspects of the images but aim at a global, signal-oriented description of the data [15].

The architecture is upgraded by image content extraction at data ingestion.

Thus, we expect to achieve a high independence of the database from user applications and to gain additional flexibility. The indexing mechanism will be enhanced by a user-oriented clustering algorithm. It adapts to user actions to improve the system responses.

3.2 Inference of Users' Conjecture Models

The general scheme to cope with this problem is clustering the images in the archive according to subjective criteria adapted to the different kinds of users. In this way we map the queries of the users, formulated in their subjective description language, to objective signal features of the image data. We use Bayesian inference [14] combined with elements of artificial intelligence to perform the mapping between subjective and objective descriptions of the image content. The general scheme is depicted in figure 3.4 [6].

In a next level of information processing the images in the archive will be clustered using conjectural models. Again, the architecture is upgraded by a

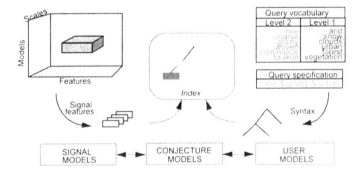

Fig. 3.4. A hierarchical modelling approach is used to link the semantic level of the query to image content. Three levels are defined: a) objective modelling of the information content of the images; this is a pure signal processing approach; b) conjecture modelling is the level which defines the adaptation to specific user applications, it makes the evidence over the model space objective; c) the user application modelling explains the semantics of the query in the context of different applications. The same images could have very different meanings in an ecological or military application. The result of these strategies in the image space is a dynamic scale/application clustering.

module covering the conjecture modelling and user adaptation. The module is included between the query engine and the relational data base system managing the archive inventory. The role of the user adaptation based on conjectural modelling is to bring the query specification process closer to the user understanding of the application, thus resulting in a different clustering of the images INDEXED IN THE INVENTORY.

As an alternative, fuzzy methods can be used for information aggregation when no models are available [6].

Finally, the application-oriented user query is answered on a semantic level. In this way, users of various fields of remote sensing applications are able to see their data in the archive.

4. Browsing Image Archives: The Role of Visual Data Mining

Prior to obtaining any images from a large image archive, a user has to specify a query. This query is based on the existing meta information stored in the archive inventory. In the following, we will call this a meta query. The user retrieves the images matching the query in a small data base (a cluster). For further details, see the left part of figure 4.1. At this stage the user has the possibility to decide which of the images resulting from the query could be suitable for the specific application. The decision is based on visual information, gained by browsing through the small image archive (figure 4.1, right part [6]).

In the case of conventional ground segments, the user browses through the quicklook images that correspond to the images matching the meta query. However, common quicklook images do not satisfy the spatial resolution re-

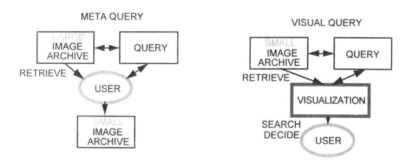

Fig. 4.1. Left - the meta query in a large image archive, the only information used is the one stored in the archive inventory, the meta information. Right - The user takes the final decision by visually browsing the image cluster resulted from the meta query.

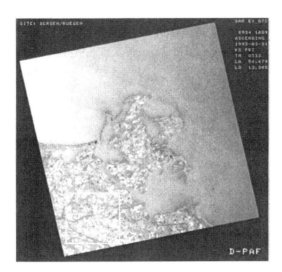

Fig. 4.2. Quicklook image of an ERS-1 GTC scene (Ruegen, Germany) as a typical example. The volume of the data is reduced by a factor of 850, the spatial resolution is reduced by a factor of 30. Thus, the information content of the quicklook image is drastically reduced. Only few image features can be distinguished: land, sea, the coastal line, and land covered by vegetation. No sufficient or adequate information for a correct selection of the interesting scenes can be obtained.

quirements of most applications. An example of an ERS-1 quicklook image is shown in figure 4.2 [6, 10].

In the case of remote sensing image data, we face an enormous volume of raw data combined with very specific analysis requests posed by the users. In order to reconcile these conflicting needs we suggest the development of a tool for interactive visual query of image products.

The interfaces of this tool will mainly consist of visual communication, i.e. they provide a sophisticated interactive "quicklook-image epitome" facility allowing the user to extract, evaluate, and visualize the significant information from high complexity multi-dimensional image data.

This exchange of information can be understood as image communication, i.e. the transfer of information via visual data. As a consequence, what we propose is an advanced remote sensing ground segment architecture designed for easier and interactive decision-making applications and broader dissemination of remote sensing data.

Example. SAR Image Epitome. Applications like SAR image transmission for ships, ice or oil slick monitoring and detection, and also ground segment applications require a new philosophy for data representation and compression. This also includes high speed and high resolution data dissemination as for example monitoring of floodings, where the transmission of high resolution data via Internet in near real time will be a major improvement. Conventional

Fig. 4.3. ERS-1 GTC scene image (Ruegen, Germany) as a test picture for the proposed compression algorithm. Left: quicklook of the full scene. Center: GTC scene, full resolution. Right: proposed quasi full resolution quicklook, compression factor 850, the same as the standard quicklook.

SAR quicklook images do not satisfy the spatial resolution requirements for such applications. As an alternative, we propose a new visual epitome based on a wavelet feature coding technique for SAR images in order to preserve the spatial resolution and to achieve high compression factors. Combining data compression, despeckling, and image restoration allows us to reach compression rates of up to about 850, thus permitting easy storage in centralized archives as well as rapid dissemination over standard networks. After decompression at the user site, the quality of the quicklook images permits the visual inspection and analysis of all spatially important image details. This becomes apparent when comparing conventional multilook quicklook images with wavelet feature coded decompressed counterparts.

Due to the extremely high compression rates, the radiometric quality of the quicklook images is degraded. However, the internal wavelet representation of the images bears the additional potential of progressive transmission that is stopped interactively when an acceptable level of radiometric fidelity is reached.

The SAR visual epitome is a quasi full spatial resolution quicklook image. The classical quicklook image generation is replaced by integrated image restoration, segmentation and compression in the wavelet transform domain (wavelet feature coding). The resulting SAR image epitome preserves almost the full spatial resolution at the same file size as the classical quicklook. The radiometric resolution is lost, but, the spatial information, extremely valuable for visual inspection, is preserved. Ships, icebergs, the demarcation of small agricultural fields, even roads and rivers are visible in the epitome image. In figure 4.3 [6] we present a typical example of images for land use and agriculture.

Due to the limited capabilities of human visual perception it becomes evident that the quality of the SAR image epitome is much better suited for visual pre-evaluation of a SAR scene than a classical quicklook image as almost all of the spatial details will be preserved.

All the upgrades mentioned before are included in the diagram of figure 4.4. Given the same volume of data and communication data rates, the user of remote sensing data will have access to the essential details of the images at a much higher visual quality during the browsing and decision process.

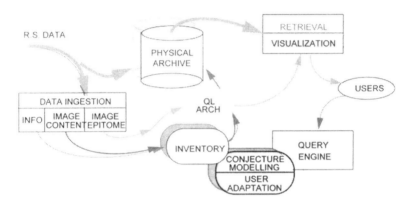

Fig. 4.4. Improving a conventional ground segment with elements of visual data mining: the image epitome as an alternative for the classical quicklook image.

The algorithm for the generation of a SAR image epitome can be integrated with a wavelet based progressive image retrieval system from remote sensing image archives. As a further development, an image enhancement and restoration step will be added in order to obtain a better visual appearance. Finally, a thematic image epitome will be developed for simultaneous and interactive visualization of collections of multi-temporal/multi-sensor data. This thematic epitome, however, calls for radiometrically and geometrically corrected images.

In the case of complex SAR data, where precise phase preservation is essential, the proposed image epitome can be modified to a quasi lossless data compression technique.

An efficient method is to apply wavelet packets to real SAR data. Therefore, we have to convert a complex data set into real format using a Hilbert transform. Conventional wavelet-based compression methods assume an energy concentration in the low frequency part. However, the Hilbert converted data sets have an energy concentration in the medium frequency band. Therefore, instead of conventional wavelet transforms, wavelet packet methods have to be used that take these spectral characteristics into account. The coefficients resulting from two iterations of a wavelet packet transform will be compressed using a linear or block adaptive quantizer. Each segment is quantized with a different number of bits ranging from zero to eight bits/sample. The bit map is based on the measured energy in each segment. The higher the energy the more bits are used. If less than 5 bits/sample are used the linear quantizer will be replaced by a block adaptive quantizer [6].

5. Summary

We presented the integration of various elements of image communication like visual data mining, query by image content, user adaptation concepts, and their integration with the architecture of conventional ground segments. The presentation was done from the perspective of information content transfer in order to remove information retrieval bottle-necks of present systems. The pragmatic goal is to give contributions for the design of the future ground segments to allow fast distribution and easy data access, near real-time applications, and to promote the implementation of electronic distribution systems.

Acknowledgement. The authors would like to acknowledge the contributions from Hubert Rehrauer and Michael Schröeder especially for making available the demonstrator. The project is supported by the ETH Research Foundation (project: 41-2637.5).

References

1. "IEEE Computer - Special Issue on Databases", September 1995.
2. "IEEE Transactions on Knowledge and Data Engineering - Special Issue on Data Mining", December 1996.
3. "IEEE Transactions on Pattern Analysis and Machine Intelligence - Special Issue on Digital Lbraries", August 1996.
4. "IEEE Computer Graphics and Applications - Special Issue on Visualization", July/August 1997.
5. "Pattern Recognition - Special Issue on Image Databases", April 1997.
6. U. Benz, M. Datcu, G. Schwarz, and K. Seidel, "Image communication: New features of a remote sensing data ground segment", in *CEOS SAR Workshop, ESTEC*, February 1998.
7. Ch. A. Bouman and M. Shapiro, "A multiscale random field model for Bayesian image segmentation", *IEEE Transactions on Image Processing*, vol. 3, pp. 162–177, December 1994.
8. M. Datcu and F. Holecz, "Generation of synthetic images for radiometric topographic correction of optical imagery", in *SPIE Proceedings on Recent Advances in Sensors, Radiometric Calibration and Processing of Remotely Sensed Data, Orlando, USA* (P. S. Chavez and R. A. Schowengerdt, eds.), vol. SPIE-1938, pp. 260–271, SPIE OE/Aerospace Sensing, 1993.

212 M. Datcu *et al.*

9. M. Datcu, K. Seidel, and M. Walessa, "Spatial information retrieval from remote sensing images–Part I: Information theoretical perspective", *IEEE Transactions on Geoscience and Remote Sensing*, vol. 36, no. 5, pp. 1431–1445, September 1998.

10. DLR-DFD, "Intelligent satellite data information system", http://isis.dlr.de/, 1998

11. G. Healy and A. Jain, "Retrieval of multispectral satellite images using physics-based invariant representations", *IEEE Transactions on Pattern Analysis and Machine Intelligence*, vol. 8, pp. 842–848, 1996.

12. D. J. C. MacKay, *Maximum Entropy and Bayesian Methods in Inverse Problems*, ch. Bayesian Interpolation, pp. 39–66, Kluwer Academic Publisher, 1991.

13. J. Mao and A. K. Jain, "Texture classification and segmentation using multiresolution simultaneous autoregressive models", *Pattern Recognition*, vol. 25, no. 2, pp. 173–188, 1992.

14. A. O'Hagan, *Bayesian Inference*, Kendall's Library of Statistics, 1994.

15. A. Pentland, R. W. Picard, and S. Sclaroff, "Photobook: Tools for content-based manipulation of image databases", in *Storage and Retrieval for Image and Video Databases II* (W. Niblack and R. C. Jain, eds.), vol. SPIE-2185, pp. 34–47, SPIE, February 1994.

16. H. Rehrauer, K. Seidel, and M. Datcu, "Multiscale markov random fields for large image datasets representation", in *Proceedings of the International Geoscience and Remote Sensing Symposium (IGARSS'97), A Scientific Vision for Sustainable Development* (T. I. Stein, ed.), vol. I, pp. 255–257, 1997.

17. C. P. Robert, *The Bayesian Choice*, vol. Springer Texts in Statistics, Berlin: Springer-Verlag, 1996.

18. M. Schröder, H. Rehrauer, K. Seidel, and M. Datcu, "Spatial information retrieval from remote sensing images: Part B. Gibbs Markov random fields", *IEEE Transactions on Geoscience and Remote Sensing*, vol. 36, no. 5, pp. 1446–1455, September 1998.

19. M. Schröder, K. Seidel, and M. Datcu, "Gibbs random field models for image content characterization", in *IEEE International Geoscience and Remote Sensing Symposium (IGARSS'97)* (T. I. Stein, ed.), vol. I, pp. 258–260, 1997.

20. M. Schröder, K. Seidel, and M. Datcu, "User-oriented content labelling in remote sensing image archives", in *Proceedings of the International Geoscience and Remote Sensing Symposium (IGARSS'98)*, 6–10 July 1998, Seattle, WA, vol. II, pp. 1019–1021, 1998.

21. K. Seidel, R. Mastropietro, and M. Datcu, "New architectures for remote sensing image archives", in *Proceedings of the International Geoscience and Remote Sensing Symposium (IGARSS'97)* (T. I. Stein, ed.), vol. I, pp. 616–618, 1997.

22. K. Seidel, M. Schröder, H. Rehrauer, and M. Datcu, "Query by image content from remote sensing archives", in *Proceedings of the International Geoscience and Remote Sensing Symposium (IGARSS'98)*, 6–10 July 1998, Seattle, WA, vol. I, pp. 393–396, 1998.

23. K. Seidel, J.-P. Therre, M. Datcu, H. Rehrauer, and M. Schröder, "Advanced query and retrieval techniques for remote sensing image databases (RSIA II and III)", Project Description Homepage, 1998, http://www.vision.ee.ethz.ch/~rsia/rsia2.html.

Fusion of Spatial and Temporal Information for Agricultural Land Use Identification – Preliminary Study for the VEGETATION Sensor

Jean-Paul Berroir

INRIA BP 105, 78153 Le Chesnay Cedex, France.
e-mail: jean-paul.berroir@inria.fr

Summary. Launched in March 1998 on the SPOT-4 platform, the VEGETATION sensor (VGT) has been designed for the analysis of vegetation dynamics. It provides daily images with 1km spatial resolution. Because of this daily sampling, it can be used for agricultural monitoring, with applications such as yield estimation and forecast, and detection of growth anomalies. These studies can furthermore be performed on large areas due to the large size of VGT scenes. The 1km spatial resolution, however, can be a problem when studying European agricultural areas: since agricultural parcels are usually relatively small, most VGT pixels contain several different land cover types and it is therefore necessary to perform pixel unmixing. For that purpose, the simultaneous use of SPOT-XS and VGT makes it possible to estimate daily reflectances of individual land coverages at the parcel level: SPOT-XS is used to provide spatial information, i.e. the location of parcels, while VGT provides temporal information at a coarse spatial scale. The estimation of agricultural parameters for large areas is achieved through operational scenarios, which depend on the amount of data available. In this chapter, we detail one scenario, making use of one SPOT image and a VGT sequence to obtain land use at high spatial resolution together with its temporal behaviour.

1. Introduction

Because of the large variety of available Earth Observation satellites, it is possible to obtain information of the same area at different spatial and temporal scales. This large amount of information can be used in various processes. Hence in the specific case of agricultural policy, satellites are expected to monitor the vegetation cycle of primary cultivations (cereals, maize, ...) at the parcel level with high temporal frequency (daily acquisitions), in order to estimate crop yields and detect anomalies such as hydric stress. In Europe however, because of the relatively small size of agricultural parcels (by comparison with the American Corn Belt), and of the high polyculture, the use of meso-scale satellite sensors (e.g. VGT) is not sufficient: they provide daily images, as required by agronomical models [2, 3, 5], but their spatial resolution (approx. 1km) is too coarse for the estimation of vegetation index at the parcel level. A way to cope with this problem is to use two different sensors: temporal (daily) information can be provided by VGT, and the fine spatial information can be extracted from SPOT-XS data.

214   J.-P. Berroir

This paper describes an original data fusion scheme, providing daily reflectances of major cultivations at the parcel level. First, a preliminary training phase, on small samples, yields the temporal profiles of reflectances and vegetation index for the main cultivations. Statistics on large areas are then estimated, with operational scenarios. Though this study is presented in the context of agricultural monitoring, it is expected that the methodology can be applied in different application contexts, to capture the spatial and temporal variability of natural phenomena.

As this study was carried out before VGT was launched, the results presented herein have been achieved using NOAA-AVHRR data instead of VGT data. Though these two sensors are quite similar, they differ in a number of important properties (see section 2.1). We expect our results to be significantly improved when VGT data are available.

This chapter is organised as follows: in section 2 we present an overview of the study; the fusion process is discussed in section 3; lastly, preliminary results of the scenario are presented and perspectives are discussed.

2. Overview of the Study

2.1 Data

SPOT-XS is used to provide accurate spatial information: its resolution is 20m, and the orbital cycle is 26 days. Images cover a 40×40 km^2 area. VGT data were not available during the study. We used instead NOAA-AVHRR data to provide the temporal information. Like VGT, the NOAA-AVHRR sensor provides daily images. Its spatial resolution is 1.1km (instead of VGT's 1km) and it provides a set of spectral channels from the visible to the thermal infra-red. NOAA-AVHRR scenes are very large, more than 1000×1000 km^2. These images are therefore well suited for studying large areas. SPOT bands 2 and 3 as well as NOAA bands 1 and 2 may be used to compute the vegetation index. It is important to notice that one encounters two main difficulties when dealing with SPOT and NOAA-AVHRR simultaneously: these two sensors don't provide the same spectral information and it is almost impossible to register their data accurately. These difficulties will vanish when VGT data are available, since SPOT and VGT are hosted on the same platform.

2.2 Learning Process

NOAA pixels (size 1.1km) are so large that they most often contain several parcels. Then subpixel modelling must be undertaken to obtain the temporal profiles of reflectances for the main cultivation observed. For that purpose, we use a training area where land use is known at high spatial resolution. We then use a classical linear mixture model [1, 4]:

$$r_i(t) = \sum_{j=1}^{N} p_{ij} \mathcal{R}_j(t) + e_i \qquad (2.1)$$

where r_i is the reflectance of NOAA pixel i, j indexes the land use within that pixel, p_{ij} is the area proportion of land use j within pixel i. $\mathcal{R}_j(t)$ is the reflectance of land-use j at date t and e_i an error term. p_{ij} are known on this training area from the land use information. The unknowns are $\mathcal{R}_j(t)$ for the main cultivation. A least square fit is then performed for each t. The output of this process is the individual temporal reflectances $\mathcal{R}_j(t)$ of each type of land use in the study area. The main assumption is that these temporal profiles are spatially constant. In other words, that agricultural practices are similar all over the study area. These profiles are used to feed agronomical models, in order to detect growth deficiencies, for instance related to hydrological stress.

2.3 Operational Scenarios

To estimate cereal production, one requires both temporal information and spatial information (i.e. that can provide the total cultivated area). The problem is to obtain the land use information over the whole site. Two operational scenarios are used for that purpose. They differ in the quality of input data.

Using only NOAA-AVHRR images, it is possible to compute meso-scale information, that is cover percentage at NOAA resolution. Each NOAA pixel i provides a reflectance time-series, and thus a set of linear relations similar to equation (2.1). The individual reflectances $\mathcal{R}_j(t)$ are known, so the inversion of the latter set of linear relations outputs the area proportion p_{ij}, that is, the land use as a percentage within meso-scale NOAA pixels.

By integrating SPOT and NOAA data through a fusion process, it is possible to estimate land use (and temporal behaviour) at the parcel level. This second scenario, will be described in the next sections. The SPOT image is first segmented into agricultural parcels. Figure 2.1 depicts an example of a SPOT-NDVI image segmentation, resulting in a set of parcels. The associated cultivations are at this point still unknown. Their identification is based on the temporal information brought by NOAA-AVHRR. For that purpose, an "energy coefficient" is minimised, which expresses the difference between observed time series and those expected according to a probabilistic framework, as explained in section 3.

3. Fusion Scheme

The fusion process aims at identifying the land-use of parcels on the basis of their temporal information, which is measured at the meso-scale by NOAA-AVHRR.

Let us consider a single NOAA pixel (i). After geometrical registration with the SPOT segmented image, we can associate it to a set of k_i parcels

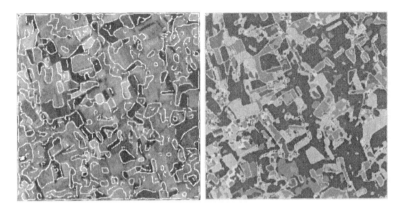

Fig. 2.1. Left: an image of NDVI superposed on contours of identified regions; Right: the result of segmentation.

whose area percentage is denoted by $s_k, k \in 1,\ldots,K_i$ (see figure 3.1). Let us associate to each parcel k the probability $P(k = n)$ that this parcel includes cultivation type $n = 1\ldots N$. The contribution of parcel k to the NOAA pixel's temporal reflectance can then be expressed as:

$$s_k \sum_{n=1}^{N} P(k = n)\, \mathcal{R}_n(t) \tag{3.1}$$

The whole pixel's reflectance r_i^m is further modelled as the sum of the contribution of each parcel it contains (equation 3.1):

$$r_i^m = \sum_{k=1}^{K_i} s_k \left[\sum_{n=1}^{N} P(k = n)\, \mathcal{R}_n(t) \right] \tag{3.2}$$

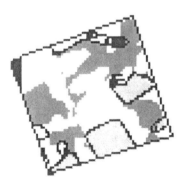

Fig. 3.1. A NOAA pixel and its component parcels.

Fusion of Spatial and Temporal Information 217

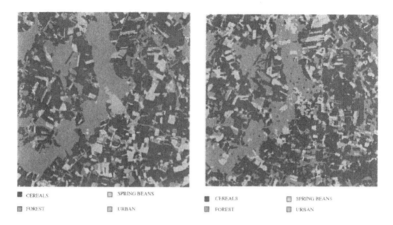

Fig. 3.2. Result of land-use identification with the fusion scheme. On this area, the main land use types are cereals, beans, forest and urban. The left image depicts the area topography, the right image the result of the fusion process.

Cultivation identification is based on the comparison between this modelled temporal behaviour $r_i^m(t)$ and the real measure $r_i(t)$. An error estimation is then defined, $e_i = \sum_t [r_i^m(t) - r_i(t)]^2$:

$$e_i = \sum_t \left[\left(\sum_{k=1}^{K_i} s_k \left[\sum_{n=1}^{N} P(k=n) \mathcal{R}_n(t) \right] \right) - r_i(t) \right]^2 \quad (3.3)$$

Equation (3.3) has to be minimised with respect to the probabilities $P(k = n)$ using a gradient method (gradient computation is straightforward) as in [6]. After computation of probabilities, each parcel is finally assigned the cultivation n maximizing $P(k = n)$. This method does not guarantee unwanted local minima, but experiments show that common cultivations (cereals, beans) and other land use (forests) are quite successfully identified (recognition rate over 80% for cereals, 60% for forests). Because of the limit in registration accuracy between the two data sources, very small parcels are never identified. We however expect significantly better results when VGT data are available, since this latter sensor is hosted on the SPOT-4 platform. Thus, the registration process will be much more accurate. Figure 3.2 shows on the right an example of our results, while the image on the left shows the ground truth.

4. Conclusion and Future Work

The proposed method is an original way to combine high precision temporal information (at coarse spatial scale) and high spatial resolution data. It has potential application to the analysis of evolving natural phenomena. Current

218   J.-P. Berroir

research includes the estimation of hydrological parameters (surface temperature, leaf area index, evapotranspiration) with both spatial and temporal variability.

The method could be improved in many ways. From a theoretical point of view, it is important to manage local minima and to define criteria making it possible to predict the performance of the process. These criteria are clearly dependent on the parcels' minimal area, and on the difference of temporal behaviour between cultivations. From the application point of view, the first improvement would be to take relief effects into account. Another difficulty is related to the assumption that agricultural practices are similar over the whole area. It would be interesting to be able to manage difference in temporal behaviour for the same cultivations at different locations.

Acknowledgement. This study has been supported by CNES under the research contract 833/CNES/94/1288/00, in collaboration with the project "Integration of VEGETATION and HRVIR data into yield estimation approach", financed by the VEGETATION program of IUC. Images have been provided by JRC-ISPRA, MARS project.

References

1. S. Bouzidi, J.-P. Berroir, and I. Herlin, "Simultaneous use of spot and NOAA-AVHRR data for vegetation monitoring", in *Proceedings of the 10th Scandinavian Conference on Image Analysis*, Lappeenranta, Finland, June 1997.
2. A. Fisher, "A model for the seasonal variations of vegetation indices in coarse resolution data and its inversion to extract crop parameters", *Remote sensing and environment*, vol. 48, pp. 220–230, 1994.
3. C. O. Justice, J. R. Townshend, B. N. Holben and C. J. Tucker, "Analysis of the phenology of global vegetation using meteorological satellite data", *International Journal of Remote Sensing*, vol. 6, pp. 1271–1318, 1985.
4. S. Mathieu, P. Leymarie, and M. Berthod, "Determination of proportions of land use blend in pixels of mutispectral satellite image", in *International Geoscience and Remote Sensing Symposium IGARSS'94*, Pasadena, CA, vol. II, pp. 1154–1156, 8–12 August 1994.
5. J. P Ormsby, B. J. Choudhury, and M. Owe, "Vegetation spatial variability and its effect on vegetation indices", *International Journal Remote Sensing*, pp. 1301–1306, 1987.
6. E. R. Panier and A. L. Tits, "On combining feasibility, descent and superlinear convergence in inequality constrained optimization", *Math Programming*, vol. 57, no. 2, pp. 261–276, 1993.

Rule-based Identification of Revision Objects in Satellite Images

Camilla Mahlander and Dan Rosenholm

Department of Geodesy and Photogrammetry,
Royal Institute of Technology, S-100 44 Stockholm, Sweden.
e-mail: millam@geomatics.kth.se

Summary. A rule based method for detection and identification of map "revision objects" in satellite data is presented. High resolution panchromatic data from the SPOT satellite are used to identify new objects in comparison to an existing map database. Medium resolution data from the wide field sensor (WiFS) of the IRS-1C satellite are tested for change detection. Fuzzy logic is used in the rules for detection and identification of the potential revision objects. Existing map data help in differentiating between new objects and objects which have been previously mapped. The development is focused on the forested parts of Sweden, and revision objects under consideration are new roads and forest "clear-cuts". Rules with different degrees of complexity are compared and the results are evaluated against visual interpretation of the satellite data. It is concluded that the rule-based method is successful for identification of clear-cuts and that the complexity of rules does not significantly affect the result. Roads are also identified, but not as successfully as clear-cuts.

1. Introduction

The increasing storage of geographical data in digital data bases is augmenting the need for cost-efficient methods for database maintenance. Among mapping organisations, the importance of database update has been widely recognised as one of the main challenges for the future and the management of this activity has evolved accordingly [9]. The methodologies for map revision are thus changing; it is no longer maps that need revision, but the databases instead. Development and implementation of cost-efficient methods for operational revision of the digital database for the topographic map of scale 1:50,000 are of high priority to the National Land Survey in Sweden (NLS). For this reason, the NLS and the Swedish National Space Board have initiated a programme for "Topographic map revision from remote sensing data". Within this programme methods for semi-automated database revision are being developed at the Department of Geodesy and Photogrammetry, Royal Institute of Technology in Stockholm.

1.1 Goal of the Study

This chapter reports on a study which has two main goals; the first is to evaluate how the complexity of rules affect the result of a rule-based method for detection and identification of revision objects in satellite data of high

220   C. Mahlander and D. Rosenholm

resolution. The second goal is to evaluate the usability of medium resolution satellite data for detection of changes with the rule-based method, which is based on fuzzy set theory.

1.2 Satellite Data for Mapping

Studies have proved that it is possible to visually detect and identify revision objects such as new roads in forest areas and new clear-cuts from Landsat Thematic Mapper (TM), SPOT XS and SPOT Pan [1, 8, 10, 11]. It has been shown that panchromatic (SPOT Pan) and multispectral (SPOT XS) satellite data are just as suitable as aerial photographs (9,200 m) for detection and identification of roads and clear-cuts [8]. SPOT panchromatic data have also been extensively used for topographic map revision in developing countries, where it has been the only economic alternative for their extensive revision needs.

In visual interpretation of satellite imagery, the interpreter uses all his experience to identify objects in the image. Not only are the grey-levels in the image considered, but also spatial and contextual factors are taken into account: size and shape of areas can be almost as important as the grey value. Knowledge based classification, where remotely sensed data are combined with other information, is considered as a possibility to mimic human visual interpretation and preliminary results from semi-automatic identification of changes in SPOT data have been promising [7, 12].

This chapter presents a rule based semi-automatic method to detect and identify changes for map revision in satellite images. Two types of changes are considered, new roads in forested areas and new clear-cuts. Data from two different satellites are used in conjunction with the existing map (in digital format) and a digital elevation model. Three steps can be identified in the revision process: change detection, identification of the revision object, and geometrical delineation of the revision object. The work presented in this chapter is focussed on the first two steps, i.e. detection and identification. The delineation phase is not dealt with here.

1.3 Fuzzy Set Theory and Image Interpretation

In visual interpretation qualitative linguistic values are frequently used. The size and shape of different objects, as well as spatial distribution of adjacent regions, etc. are factors which are considered as complementary to spectral characteristics. These factors often include vague and ill-defined natural language such as *a big area* or *a steep slope*. Fuzzy sets are well suited for translating these linguistic values into mathematical terms. It is appropriate to use fuzzy sets when ambiguity and vagueness in mathematical or conceptual models of empirical phenomena are dealt with [3, 15]. Altman [2] uses the underlying assumption in his work that conversion of imprecise data to 'hard' data introduces error at too early a stage of the analysis process. For example, before performing a buffer analysis. the relationship of being near to a river would be translated into one of being less than 5 km from the river. The sudden transition at the buffer boundary does not necessarily correspond with our notion of nearness as involving gradual change. By deferring the hardening process to the final stage of the analysis a more accurate result is expected. The most important advantage with fuzzy sets is perhaps the ability to express the amount of ambiguity in human thinking and subjectivity (including natural language) in a comparatively undistorted manner [13]. In the field of remote sensing, fuzzy sets have been used for image interpretation and image classification (e.g. [5, 14]).

1.4 Data Availability

The availability of remotely sensed data is improving. Both spatial and spectral resolution are increasing and new satellites are being launched with a much higher frequency than in the past [4]. Currently available satellite data suitable for mapping have a maximum resolution of approximately 5 metres. Commercial satellites with spatial resolution of up to 1 metres are, however, in preparation. It is believed that revision by satellite imagery will become less costly than mapping by aerial photography and can lead to much faster results [6].

In this study two types of satellite data were used, a SPOT Panchromatic image and WiFS data from IRS-1C.

2. Methods

2.1 Change Detection

Recent clear-cuts and roads contrast strongly with the surrounding forest in imagery and change detection can be based on spectral characteristics alone, focusing on spectral differences. However, to distinguish between roads, clear- cuts, and rock-outcrops (for example) the spectral differences are not sufficient as they all have similar spectral characteristics.

The change detection process is based on comparison between the existing topographic map in digital form and satellite images. The map data are used to mask out areas which are of no interest for a certain change, or to exclude object types which could easily be mixed with new roads or clear-cuts. For example, all areas which have not previously been mapped as forest are excluded when clear-cuts are identified. Rock-outcrops, marshes and power-line corridors, which could be mistaken for clear-cuts, are also excluded when necessary. Digital elevation data are included in the data base and used to exclude very steep areas where the vegetation usually is sparse. since these areas might cause confusion with clear-cuts.

2.2 Study Area and Data

The study area is located in central Sweden and covers an area of 625 km² which corresponds to a topographic map sheet (1:50,000). The area is almost entirely covered by coniferous forest and logging is intensive. Approximately 270 unmapped clear-cuts of different size and about 55 new roads are found within the area.

The SPOT panchromatic scene was recorded on August 13, 1995. The IRS-1C WiFS scene was recorded on September 15, 1996. WiFS is optimised for the collection of vegetation indices and has two bands, red and short-wave infrared.

2.3 Rule Based Identification in SPOT Pan

A set of rules was developed for detection and identification of changes. The rules are based on spectral, contextual and spatial characteristics of the revision objects in order to cover some of the factors a person performing a visual interpretation would consider. Since the changes of interest are well defined and only of two types, the rules can be created in a specific way. Rules with varying complexity have been used and the results have been compared. The most complex rules are shown below. The complexity may, for example, vary according to the available data.

Rule for identification of clear-cuts (rule 1):

1.	IF	forest-area
2.	ANDIF	high-spectral-reflectance
3.	ANDIF	larger-than-minimum-mapping-unit
4.	ANDIFNOT	marsh
5.	ANDIFNOT	powerline-corridor
6.	ANDIFNOT	road
7.	ANDIFNOT	too-steep
8.	ANDIFNOT	too-far-from-existing-road/clear-cut
	THEN	new-clear-cut

Rule for identification of roads (rule 2):

1.	IF	high-spectral-reflectance
2.	ANDIF	line-shaped-and-thin
3.	ANDIF	connected-to-mapped-road
4.	ANDIF	larger-than-minimum-mapping-unit
5.	ANDIFNOT	previously-mapped-as-road
	THEN	New-road

For some of the factors in the above rules the topographic map can easily be used to mask out areas with no interest. However, other factors involve

Rule-based Identification of Revision Objects in Satellite Images   223

vague statements which makes it impossible to just use a mask. In the above rules *high-spectral-reflectance, too-steep, too-far-from-existing-road/clear-cut* and *line-shaped-and-thin* are examples of statements which include ill-defined expressions. To translate and combine these vague statements into mathematical terms fuzzy logic was used. All factors in the rules were then represented as raster layers which were combined, and a resulting layer showing potential clear-cuts/roads was obtained. The result was compared to a visual interpretation which was based on the SPOT image in conjunction with print-outs from Landsat TM colour composites and field visits.

2.4 Change Detection in WiFS

One of the main advantages of using medium resolution data such as WiFS is that a large area is covered by one scene. If this kind of data proves to be useful for change detection it could be used as the first step in the revision process, indicating on which areas to concentrate further revision work. It is believed to be of great economical importance if data with high coverage could be used operationally. The spectral information from the WiFS data could also be used in combination with panchromatic data with higher spatial resolution, such as SPOT Pan.

In order to evaluate how useful WiFS data are for detection of changes, the rule based method was used also for this scene. The possibility of whether this data type could be used to detect changes for map revision purposes was investigated. The focus in this investigation was thus on detection and not identification, but it was assumed that most detected changes refer to new clear-cuts. Rules were constructed in the same way as for SPOT data. Instead of high-spectral-reflectance, however, a normalised vegetation index (NDVI) based on the two WiFS bands (red and short wave infrared) was used.

Rule for change detection in WiFS (rule 3):

1.	IF	forest-area
2.	ANDIF	low-NDVI
3.	ANDIF	larger-than-minimum-mapping-unit
4.	ANDIFNOT	marsh
5.	ANDIFNOT	powerline-corridor
	THEN	change (potential clear-cut)

3. Test and Evaluation of the Rules

In order to evaluate the results from the semi-automatic identification process a visual interpretation of changes in the test area was performed. Typical areas which were representative for different classes were also visited in the field. The documentation from the field work served as a basis for the visual interpretation. The SPOT image was used in conjunction with multispectral Landsat TM data and field information and new clear-cuts and roads were digitized on-screen. Since the WiFS scene was recorded a year later than the SPOT scene, it included a somewhat higher number of revision objects. The influence of the different factors in the rules was tested by first using a simple rule, including only a few statements, and then increasing the complexity by adding other factors one by one. Each result from the semi-automatic interpretation was compared to the result from the visual interpretation. The evaluation of the results was based on regions instead of pixels.

A combination of SPOT and WiFS data was also tested for identification. For this test one of the rules which had given a satisfying result for clear-cut identification in SPOT was used in conjunction with the NDVI derived from the WiFS scene. An evaluation was carried out to see if this combination of data sets would increase the accuracy of identification.

4. Results

The results of identification of clear-cuts and roads are found in tables 4.2 and 4.4, and the result of change detection in WiFS is found in table 4.6. The figures for included factors refer to the numbers in rules 1–3. Tables 4.1, 4.3 and 4.5 include the evaluation criteria. Error of omission corresponds to the percentage of clear-cuts/roads which have been wrongly omitted, error of commission refers to the percentage of areas which have been falsely identified as clear- cuts/roads. In both cases the results are given in relation to the size of clear-cuts and roads identified in the visual interpretation. A clear-cut or road is treated as a unit which can not be partly identified. The entire clear- cut area is considered as identified even if the geographical position of the semi-automatically identified clear-cut does not fully correspond to the visually identified area. There is thus not an error of omission if only a part of a clear-cut is missing. Accordingly, areas which are falsely identified as clear-cuts are not included in the error of commission if the main part of the areas are connected to an unmapped clear-cut. Roads are treated in the same way as clear-cuts, and for practical reasons the size (expressed in hectares) instead of the length of the roads is used in the evaluation.

4.1 Clear-cuts Identified from SPOT Data

The results in table 4.2 show that all rules involving only SPOT data correctly identify 89-92% of the clear-cut areas when all clear-cuts, independent of size, are considered. The largest clear-cuts were most successfully identified. Clear-cuts larger than 5 hectares were correctly identified to a degree of 96%, those larger than 10 hectares to 98% and clear-cuts larger than 15 hectares were identified to a degree of 100%. The complexity of the rules

Rule-based Identification of Revision Objects in Satellite Images   225

Table 4.1. Evaluation criteria for identification of clear-cuts in SPOT data.

A	Correct identification, all included	F	Error of omission, all included.
B	Correct identification, > 3 hectares.	G	Error of omission, > 3 hectares.
C	Correct identification, > 5 hectares.	H	Error of omission, > 5 hectares.
D	Correct identification, > 10 hectares.	I	Error of omission, > 10 hectares.
E	Correct identification, > 15 hectares.	J	Error of omission, > 15 hectares.
-		K	Error of commission.

Table 4.2. Semi-automatic identification of clear-cuts in SPOT Pan based on rule 1. Results expressed in per cent of total area according to visual interpretation.

Included factors.	A	B	C	D	E	F	G	H	I	J	K
all	89	93	96	98	100	11	7	4	2	-	11
1–7	91	95	96	98	100	9	5	4	2	-	13
1–6	91	95	96	98	100	9	5	4	2	-	13
1–2, 4–6	92	95	96	98	100	8	5	4	2	-	15
1–2, 4–5	92	95	96	98	100	8	5	4	2	-	18
1–6, NDVI from WiFS	64	72	81	95	100	36	28	19	5	-	0.2

affects mainly the smaller clear-cuts. Complex rules cause omission of correct areas to a larger extent than simple rules. The complexity also affects the error of commission: the higher the complexity, the lower is the error of commission. The difference is small: the simplest rule gave an error of commission of 18% and the most complex an error of commission of 11%. A correlation between error of commission and error of omission for small clear-cuts exists. A high complexity gives a smaller error of commission and a higher error of omission and vice versa.

The most important factor for identification of clear-cuts is the spectral reflectance in combination with a forest mask. Also the rule for exclusion of new roads is important, without this rule the error of commission is as high as 18%. The size of the clear-cuts is important, and the inclusion of size limit in the identification rule lowered the error of commission by 2%. At the same time however, the error of omission increased by 1%. The statement concerning slope did not affect the result at all, while inclusion of distance to existing roads/clear-cuts lowered the error of commission and increased the error of omission by 2%.

4.2 Clear-cuts Identified from SPOT and WiFS Data

The inclusion of spectral information from the WiFS scene significantly lowered the error of commission. When only SPOT data were used it was between 11% and 13%, while inclusion of WiFS data decreased the error of commission to 0.2%. However, the error of omission was as high as 36% for smaller clear-cuts, while those with a size of 10 hectares or more were successfully identified with errors of omission of 5% or less.

4.3 Identification of Roads

The identification of roads was not as successful as the identification of clear-cuts. In the best result 85% of the road areas were correctly identified. The error of commission was 15%. The most complex rule gave an error of commission of 11%, but only 65% of the road areas were correctly identified. The shape of the road is the most important factor and it identifies 88% of the roads. The shape has to be used in conjunction with spectral reflectance, otherwise the commission of other areas is very high (121% in this case).

Table 4.3. Evaluation criteria for identification of roads in SPOT data.

A	Correct identification, all included	D	Error of omission, all included.
B	Correct identification, > 0.5 hectares.	E	Error of omission, > 0.5 hectares.
C	Correct identification, > 1 hectares.	F	Error of omission, > 1 hectares.
G	Error of commission.		

Table 4.4. Semi-automatic identification of roads in SPOT Pan based on rule 2. Results expressed in per cent of total area according to visual interpretation.

Included factors.	A	B	C	D	E	F	G
all	63	65	69	37	35	31	0.3
1–3, 5	65	66	69	35	34	31	11
1–2,4–5	83	85	89	17	15	11	14
1–2, 5	85	87	92	15	13	8	15
2	88	90	95	12	10	5	121

4.4 Change Detection in WiFS

The test of change detection in WiFS imagery was based on a simple rule. Since the more complex rules did not significantly improve the results for identification in SPOT images it was assumed that the change detection with WiFS would not be improved if the factors for distance to existing clear-cut/road and slope were included. The result of the change detection was not very successful: the error of omission varied between 22% and 55% and the error of commission was 48%. The lower figures for error of omission refer to large areas of changes.

Table 4.5. Evaluation criteria for change detection in WiFS.

A	Correct detection, all areas	E	Error of omission, all areas.
B	Correct detection, > 5 hectares.	F	Error of omission, > 5 hectares.
C	Correct detection, > 10 hectares.	G	Error of omission, > 10 hectares.
D	Correct detection, > 15 hectares.	H	Error of omission, > 15 hectares.
I	Error of commission.		

Table 4.6. Semi-automatic detection of changes in WiFS based on rule 3. Results expressed in per cent of total area according to visual interpretation.

Included factors.	A	B	C	D	E	F	G	H	I
1–5	45	63	80	78	55	37	20	22	48

5. Discussion and Conclusions

It is concluded that SPOT data can be successfully used for semi-automatic identification of new clear-cuts. The rule-based method works best on large revision objects. Increased complexity of rules did not significantly improve the result. The satellite image and the existing topographic map are sufficient for identification. In order to decrease the error of commission, the vegetation index derived from the WiFS image proved to be useful.

Change detection based on WiFS data only was not very successful. Both error of commission and error of omission were unacceptably high. The low resolution and geometrical problems are probably the main reasons for the poor result. It was obvious, especially around the lakes, that the satellite image and the mask did not overlap properly. Furthermore, only areas with low values of NDVI were extracted and there is a risk that only areas corresponding to the most recent clear-cuts are identified in this way. Older clear-cuts are usually covered with vegetation, and the NDVI is thus higher. It is believed that the full potential of the WiFS image has not yet been evaluated, further development and tests are needed for these data.

The results for road identification are not yet satisfactory. The inclusion of size thresholding in the rule did not improve the result in the expected way. A visual evaluation of the result indicated that the main reason for this is that not all of the identified road pixels are fully connected -correct areas might thus be omitted since the region they define is smaller than the size threshold. The same explanation is valid for the rule which includes the statement that potential road areas should be connected to existing roads. Most of the larger and longer roads were successfully identified, while the smaller ones were omitted.

The shape of the roads is clearly an important factor and filters for identification of elongated, thin areas are thus of high importance. Improved filters could probably give better results. The possibility of handling gaps in potential roads also has to be incorporated in the methodology. Once these problems are solved it is likely that the results will be improved.

The major advantage of using fuzzy logic in the rules is the ability to formulate "soft" decision rules. No "hard" decisions are made early in the analysis process, and the risk of omitting pixels that belong to a revision object is reduced. Also, in the final phase of the revision process, when the revision objects are to be delineated, the use of fuzzy logic can also be expected to be advantageous. Connected pixels with varying membership values can be allowed to form an object. The possibilities of deriving accurate object

228   C. Mahlander and D. Rosenholm

boundaries are thought to increase this way. Iterative changes of the borders are possible, based on different degrees of membership.

Acknowledgement. This project has been carried out within the programme for Topographic Map Revision from Remote Sensing Data, initiated by the National Land Survey of Sweden, and the Swedish National Space Board. The latter is financially supporting the project.

References

1. F. Ahern and D. Leckie, "Digital remote sensing for forestry: requirements and capabilities, today and tomorrow", *Geocarto International*, vol. 2, no. 3, pp. 43–52, 1987.
2. D. Altman, "Fuzzy set theoretic approaches for handling imprecision in spatial analysis", *International Journal of Geographical Information Systems*, vol. 8 no. 3, pp. 271–289, 1994.
3. P. A. Burrough, "Fuzzy mathematical methods for soil survey and land evaluation", *Journal of Soil Science*, vol. 40, pp. 477–490, 1989.
4. L. W. Fritz, "The era of commercial Earth observation satellites", *Photogrammetric Engineering and Remote Sensing*, vol. 62 no. 1, pp. 39–45, 1996.
5. S. Gopal and C. Woodcock, "Theory and Methods for Accuracy Assessment of Thematic Maps Using Fuzzy Sets", *Photogrammetric Engineering and Remote Sensing*, vol. 60 no. 2, pp. 181–188, 1994.
6. G. Konecny, "International Mapping from Space", *International Archives of Photogrammetry and Remote Sensing*, vol. XXXI (B4), pp. 465–468, 1996.
7. C. Mahlander, D. Rosenholm and K. Johnsson, "Semi-automatic identification of revision objects in high resolution satellite data", *International Archives of Photogrammetry and Remote Sensing*, vol. XXXI, (B4), pp. 534–539, 1996.
8. B. Malmström and A. Engberg, "Evaluation of SPOT data for Topographic Map Revision at the National Land Survey of Sweden", *International archives of photogrammetry and remote sensing*, vol. XXIX (B4), pp. 557–562, 1992.
9. P. Newby, "Working group IV/3 report and review of progress in map and database revision", *International Archives of Photogrammetry and Remote Sensing*, vol. XXXI (B4), pp. 598–603, 1996.
10. H. Olsson and J. Ericsson, "Interpretation and Segmentation of Changed Forest Stands from Difference Imagery, Based on Regression Function", in *Proceedings, Central Symposium of the International Space Year Conference (ESA SP-341)*, Munich, Germany, March 30 – April 4, 1992, pp. 761– 765, 1992.
11. P. Pilon and R. Wiart, "Operational Forest Inventory Applications Using Landsat TM Data: the British Columbia Experience", *Geocarto International*, vol. 5, pp. 25–30, 1990.
12. R. Solberg, "Semi-Automatic Revision of Topographic maps from Satellite Imagery", *International archives of photogrammetry and remote sensing*, vol. XXIX (B4), pp. 549–556, 1992.
13. T. Terano, K. Asai and M. Sugeno, *Fuzzy systems theory and its applications*, Academic press, 1992.
14. F. Wang, "Towards a Natural Language User Interface: an Approach of Fuzzy Query", *International Journal of Geographic Information Systems*, vol. 8 no. 2, pp. 143–162, 1994.
15. L. Zadeh, "Fuzzy Sets", *Information and Control*, no. 8, pp. 338–353, 1965.

Land Cover Mapping from Optical Satellite Images Employing Subpixel Segmentation and Radiometric Calibration*

Werner Schneider

Institute of Surveying, Remote Sensing and Land Information (IVFL)
Universität für Bodenkultur (BOKU, University of Agricultural Sciences)
Peter-Jordan-Straße 82, A-1190 Vienna, Austria.
e-mail: schneiwe@mail.boku.ac.at

Summary. A general trend in remote sensing image analysis can be observed today away from multispectral methods towards structural analysis. Multispectral information is, however, indispensable for land cover identification in agriculture, forestry and the natural environment. This contribution tries to point out the synergetic effects to be achieved by combined use of multispectral and of structural concepts. Spatial subpixel analysis as a preprocessing step for segmentation and multispectral classification is described. A procedure for combined subpixel-fine segmentation and classification is presented. Possibilities for radiometric calibration for enabling the use of general spectral knowledge sources are proposed.

1. Introduction: Spectro-Radiometric Versus Structural Analysis

This chapter addresses the use of *computer vision methods* in remote sensing, while at the same time accentuating *radiometric aspects* of the images.

To portray image analysis in simplified terms, one can distinguish two different ways of looking at an image:

- "The physicist's point of view": An image is seen as a huge set of (tens or hundreds of millions of) radiometric measurements. Although these measurements are not completely independent of each other, with spatial correlations between them, they are individual measurements. Each individual measurement may contain a wealth of spectro-radiometric information [1].
- "The point of view of computer vision": An image is seen as a set of information entities which are very similar to those we collect with our eyes in every-day life. Emphasis is placed on the structure (spatial distribution) of image details, on the outline of objects to be recognized in the image. Radiometry is not of major concern, it is only taken into account in relative terms.

---

* This work was financed by the Austrian "Fonds zur Förderung der wissenschaftlichen Forschung" (project S7003).

Both approaches are important in remote sensing. Depending on the field of application, the one or the other approach may be to the fore. Spectro-radiometric information is indispensable for land cover identification in agriculture, forestry and the natural environment. Structural information is essential for the identification of man-made objects, especially in urban areas.

A general shift in remote sensing image analysis can be observed today from multispectral methods to structural analysis. One reason for this is that, from an applications point of view, urban mapping and monitoring is gaining importance at the expense of environmental applications. Another point is that multispectral information has been put to use mainly in conventional pixelwise statistical classification. From the poor performance or even failure of these conventional methods it is sometimes erroneously concluded that multispectral information is of limited usefulness and reliability. However, the usual problems with multispectral analysis (e.g. the fact that results are not reproducible and that methods cannot be generalized) mainly stem from the purely statistical use of uncalibrated spectro-radiometric data.

This contribution tries to emphasize the *continuing significance of spectro-radiometric information* in remote sensing and the *synergy to be obtained by combined use of multispectral and structural concepts*. In particular,

- the improvements in spectro-radiometric analysis by spatial subpixel analysis (alleviating the mixed pixel problem),
- a procedure for combined subpixel-fine segmentation and classification and
- possibilities of radiometric calibration for enabling the use of general spectral knowledge sources

are discussed.

2. High Spatial Resolution Land Cover Mapping

The main problem to be solved can be stated as follows: Detailed spatial land use patterns prevail in most parts of Europe. Therefore, a high spatial resolution of remotely sensed images for land cover mapping is required. While the new high resolution sensors available today and announced for the near future may offer adequate spatial resolution, they do not provide the high spectral and radiometric resolution necessary for sufficient thematic land cover differentiation. Therefore, "conventional" multispectral sensors such as Landsat TM will be needed also in the future.

Two approaches can be followed: One can either try to obtain high-spatial-resolution land cover information from medium-spatial- resolution imagery by spatial subpixel segmentation, or one can fuse information from a sensor with high spatial resolution and a second sensor with high spectral and high radiometric resolution.

This contribution concentrates on the former approach. Spatial subpixel segmentation can also be used as a preprocessing step for subsequent object-based fusion of information from both types of sensors.

3. Spatial Subpixel Analysis

There are two approaches to subpixel analysis:

In *spectral subpixel analysis*, a-priori knowledge on the categories of land cover in the image and on the spectral characteristics of these categories is assumed. One tries to describe every pixel as a linear combination of the mean pixels of these categories. The method gives no information on the spatial distribution within one pixel.

In *spatial subpixel analysis*, the spatial distribution of pixel values in the neighbourhood of a pixel is used to determine the spatial distribution *inside* the pixel. A geometric model of this spatial distribution is assumed, and the parameters of the model are estimated from the pixel values in the neighbourhood.

In the method described in [2] every pixel is analyzed within the context of its 8 neighbouring pixels. Certain models of the scene pattern within this cell of 3×3 pixels may be assumed. The simplest model is shown in figure 3.1. The scene pattern consists of 2 homogeneous regions separated by a straight boundary. Four "subpixel parameters" are necessary to describe this model: two geometric parameters specifying the position of the boundary line, and two radiometric parameters specifying the "pure pixel" values p_1 and p_2 within the homogeneous regions. The geometric parameters are the angle θ defining the orientation of the boundary line, and the distance d of the boundary line from the centre.

More complicated models can be defined. The maximum number of parameters of any model must not exceed 8 for a cell of 9 pixels. For every model, the parameters are estimated from the 9 original pixel values. The

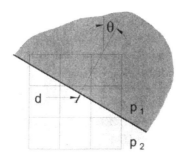

Fig. 3.1. Simple model for spatial subpixel analysis: geometric parameters θ, d, and radiometric parameters (pure pixel values) p_1, p_2.

Fig. 3.2. Synthetic input image (left) and resulting output image after subpixel analysis and resampling to 1/3 of original pixel size (right). Only the model shown in figure 3.1 has been used. Therefore, the pixels of the original image close to corners are not resolved to subpixel detail.

sum of squares of the residuals is a criterion for the appropriateness of a model. The model most appropriate in this sense is finally selected, and the subpixel parameters for this pixel are stored. If no model with residuals below a certain threshold can be found, the pixel is labelled as one without subpixel information.

The estimation of subpixel parameters is complicated. To speed up computation, various algorithmic improvements have been implemented [3].

The resulting subpixel parameters can be used to resample the original image with a smaller pixel size. Figure 3.2 shows a synthetic input image together with the result obtained by spatial subpixel analysis using the simple model of figure 3.1 and resampling to 1/3 of the original pixel size. Figure 3.3 shows a Landsat TM image after the same treatment. The result has a pixel size of 10 m.

Fig. 3.3. Original Landsat TM (band 3) image (left) and resulting output image after subpixel analysis and resampling to 1/3 of the original pixel size, i.e. to 10 m pixel size (right).

This method of spatial subpixel analysis is not to be confused with edge enhancement techniques. The main aim is not to find the exact geometric position of the boundary line, but to solve the mixed pixel problem by finding the pure pixel values and thereby facilitate multispectral analysis.

4. Image Segmentation and Classification

Classical pixelwise classification suffers from the problems of

– mixed pixels
– edge effects influencing texture classification
– ignoring shape information.

In an effort to simulate the working method of a human interpreter, image segmentation is employed in addition to classification. The purpose of segmentation is to obtain image regions with a meaning in the real world. In the case of land cover classification, the regions should represent parcels of agricultural land, forest stands, villages, rivers, lakes etc. The problem is the interdependence of segmentation and classification: Classification results are needed as input for a meaningful segmentation, and the segmentation results are required for a good classification (e.g. in order to be able to use texture and shape parameters).

The following method is applied for land cover mapping:

step 1: pixelwise classification: The original Landsat TM image data are classified pixel-by-pixel according to the maximum likelihood method. For every pixel, in addition to the classification result in form of a category index, a-posteriori probabilities for every category are generated. The problems of this step are
  – the definition of useful sub-categories with normally distributed multispectral signatures and
  – the provision of sufficient and reliable training data.
step 2: subpixel analysis: This step is described in detail in the previous chapter. The result is subpixel information for certain pixels for which an appropriate spatial subpixel model has been found.
step 3: segmentation: Segmentation is performed according to the region growing method. Seed pixels may be placed in the image randomly or at the centroids of homogeneous clusters of categories obtained by classification in step 1. Regions grow as long as homogeneity conditions in terms of multispectral data and of category probability densities are satisfied. In addition, termination conditions may be defined for subpixel boundaries. The resulting regions are represented in vector form with boundaries at subpixel resolution. Figure 4.1 gives an example of the result of a segmentation of a Landsat TM image from an agricultural area in the eastern part of Austria (Burgenland).

Fig. 4.1. Segmented Landsat TM image of agricultural area.

step 4: regionwise classification: Finally, the regions are assigned to categories by decision tree classification. Spectral, textural and shape parameters (including subpixel parameters) as well as pixelwise category probability densities are used as features for the classification. Figure 4.2 shows a result of segment-wise landcover classification from an area at the river Danube in Upper Austria.

Fig. 4.2. Result of segment-wise land cover classification.

5. Automated Radiometric Calibration of Remotely Sensed Images

The procedure described so far can only be applied to small areas. If the area to be mapped covers more than one image, a complete set of training data (ground truth) for every image has to be provided in step 1 of the above procedure, and different multispectral parameters for decision tree classification have to be used in step 4 in order to convey the necessary radiometric characteristics of every land cover category.

Alternatively, the images may be radiometrically calibrated. In this case, a constant knowledge base (training data set, decision tree parameters) can be used for the entire set of images.

The concept of radiometric calibration is illustrated in figure 5.1. Every land cover type is characterized by reflectance values ρ on the terrain. The corresponding pixel values d in the image, which should be a measure of these terrain reflectances, are influenced by disturbances, mainly by the atmosphere. Radiometric calibration of an image denotes the establishment of the transformation between pixel values in the image and reflectance values on the terrain.

Radiometric calibration is difficult if not impossible to implement in practice if external parameters describing the state of the atmosphere at the time of image acquisition are required. In the optimal case, it should be possible to derive the entire information for calibration from the image itself.

The basic idea of an approach for this type of in-scene radiometric calibration of remotely sensed images is outlined in figure 5.2. In the terminology used here, a distinction is made between image objects (in the image, with

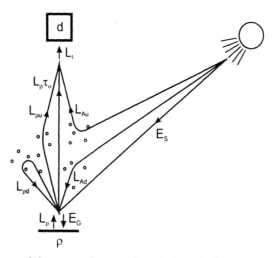

Fig. 5.1. Influence of the atmosphere on the relationship between terrain reflectance ρ and pixel value d.

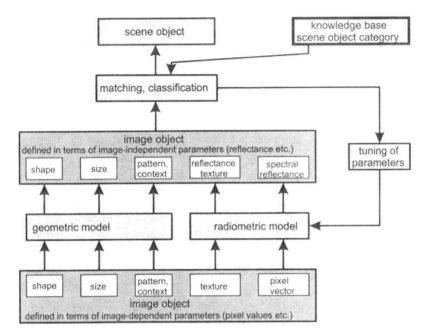

Fig. 5.2. Simultaneous classification and radiometric calibration.

remote-sensing-specific attributes such as pixel values or reflectance values) and scene objects (on the terrain, with application-specific attributes such as land cover). Image objects are obtained by segmentation of an image. Firstly, the various attributes of the image objects such as spectral characteristics (mean pixel vector) are image-dependent: The same object may have different attributes on two images (e.g. acquired at different dates). The *image-dependent* attributes are converted to *image-independent* attributes by transformations making use of a *geometric model* and a *radiometric model* of image acquisition (*georeferencing* and *radiometric calibration*). The transition to scene objects with application-specific attributes is then accomplished by classification, which is a matching with scene object prototypes in a knowledge base.

The weak point of the procedure is the radiometric model. For operational applications, the (image-dependent) parameters of this model should be derived from the image itself. This is done by *simultaneous classification and radiometric model parameter estimation*: Starting from plausible initial values, the radiometric model parameters are varied systematically until a maximum match between image objects and prototype objects in the knowledge base is reached.

It should be pointed out that, in this unified method, comprehensive use of spectro-radiometric as well as of structural image information is made, both for radiometric calibration and for thematic image analysis. For example, shape properties of image objects may influence the radiometric parameters: If an image object, due to its rectangular shape with straight boundaries, is more likely to represent an agricultural field than a forest, this may have some bearing on the radiometric calibration.

At present, the potential and the limitations of this method are being evaluated.

6. Conclusion and Outlook

Spectro-radiometric information is important in remote sensing in certain fields of application, especially for mapping and monitoring the natural environment. Some of the difficulties encountered in the past with multispectral classification of remotely sensed images can be alleviated or overcome by employing computer vision concepts of structural analysis, such as subpixel analysis and image segmentation. A procedure of combined subpixel segmentation and classification for land cover mapping from medium-resolution images such as Landsat TM has been developed. The method produces satisfactory, detailed land cover maps. It is not possible to obtain land cover maps of a similar quality from high resolution images available at the moment and announced for the near future, as the multispectral capabilities and the radiometric quality of these images does not provide sufficient thematic information on land cover in a vegetation-dominated environment.

Additional improvements can be expected for the future from the combined use of high-spatial-resolution pan images and medium-spatial-resolution high-spectro-radiometric-quality images.

References

1. J. S. MacDonald, "From space data to information", in: *Proceedings International Society for Photogrammetry and Remote Sensing, Joint Workshop "Sensors and Mapping from Space" of Working Groups I/1, I/3 and IV/4*, October 2, 1997, Hannover, Germany, Institute for Photogrammetry and Engineering Surveys, University of Hannover, vol. 17, pp. 233–240, 1997.
2. W. Schneider, "Land use mapping with subpixel accuracy from Landsat TM image data", in: *Proceedings 25th International Symposium on Remote Sensing and Global Environmental Change*, Graz, Austria, 4-8 April 1993, pp. II-155–II-161, 1993.
3. J. Steinwendner and W. Schneider, "Algorithmic improvements in spatial subpixel analysis of remote sensing images", in: *Pattern Recognition and Medical Computer Vision 1998, Proceedings of the 22nd Workshop of the Austrian Association for Pattern Recognition, Österreichische Computer Gesellschaft*, pp. 205–213, 1998.

Semi-Automatic Analysis of High-Resolution Satellite Images

Wim Mees and Marc Acheroy

Royal Military Academy, Signal & Image Centre,
Renaissancelaan 30, B-1000 Brussels, Belgium.
e-mail: {wim,acheroy}@elec.rma.ac.be

Summary. In this chapter we present a semi-automatic scene analysis system. The image interpretation knowledge that must be integrated in the scene analysis system in order to regularize the otherwise ill-posed scene analysis problem, its representation and integration will be analyzed. Finally, the results of a prototype system will be presented.

1. Introduction

In a study [21] performed by the *TNO Physics and Electronics Laboratory* and the *National Aerospace Laboratory* in the Netherlands, the authors found that the manual extraction of cartographic information from remote sensing images requires resolutions of 5 metre or less. A feasibility study [1], performed by the Belgian *Signal & Image Centre* for the *West European Union Satellite Centre*, reports similar resolution requirements for automatic feature extraction algorithms. Therefore the advent of a new generation of commercial high-resolution earth observation satellites, producing 1m panchromatic and 5m multi-spectral imagery, offers interesting possibilities for information gathering.

Some authors, such as Tannous *et al.* in [20], suggest that because of the lack of maturity of fully automatic extraction tools, the most promising way of increasing the productivity of the image analysis process consists in requiring the user to provide the initial positioning of the primitives of interest and thereafter perform the fine adjustment of these primitives to the images by means of automatic image processing algorithms. In this chapter we will describe an approach that, given the limitations of current automatic object extraction algorithms, nevertheless results in a semi-automatic scene analysis system which only requires a limited user intervention.

2. Semi-Automatic Scene Analysis

The inverse problem of aerial image interpretation is an ill-posed problem [18]. Its regularization requires the application of a priori knowledge about the scene, constraints expressing general knowledge about the real-world environment and the use of a limited, discrete solution space [6]. This may seem like a harsh restriction at first but this isn't really so. Indeed, our real-world environment is highly structured and is constrained by human habits and physical and administrative laws to a number of basic patterns. Its apparent complexity is produced from this limited vocabulary of basic forms by compounding them in different combinations. If the intrinsic complexity of our environment were approximately the same as its apparent complexity, there would be no lawful relations and no intelligent prediction. It is the internal structuring of our environment that allows us to reason about it successfully using the simplified descriptions we typically employ [17].

Image interpretation knowledge is qualitative knowledge [4]: it doesn't always allow for a correct, consistent and complete match between the represented domain knowledge and the real-world situation, but it can nevertheless be used to get an approximate characterization of the behaviour of the modeled domain.

All scene analysis systems apply this qualitative image interpretation knowledge in order to solve the otherwise ill-posed scene analysis problem. The way in which they represent and integrate it, differs from one system to another.

One of the first systems for the automatic analysis of aerial images was called *"Structures Analysis of Complex Aerial Photographs"* (SACAP), developed by Nagao *et al.* [13]. It consists of a blackboard system, on which the regions, resulting from a segmentation of the input images, are posted. A series of knowledge sources classify these regions based on their radiometric and geometric characteristics. Alongside the knowledge sources. there is also a control system linked to the blackboard which solves conflicts (a single region is assigned two different classifications) and removes indecision (a region is not assigned a class at all).

At Stanford University, Brooks [2, 3] developed the ACRONYM system for the detection of 3D objects in 2D images, applied to the interpretation of airplanes on aerial images. The system consists of a constraint manipulation module and a prediction module.

Mc Keown *et al.* [7] developed a *"System for Photo Interpretation of Airports using Maps"* (SPAM) using aerial images of airports. It was later extended to suburban areas [8]. The system combines rules and frames as knowledge representation methods.

The scene analysis and 3D modeling system AIDA [5], developed at the *Institut für Theoretische Nachrichtentechnik und Informationsverarbeitung* of the *Universität Hannover*, uses for its image understanding part, a semantic network [22], implemented in a frame-like structure [19]. This network consists of concept nodes which describe a generic world-model, instance nodes corresponding to objects found in the scene and different types of links between the nodes. The AIDA system uses a graph processing algorithm for instantiating nodes and links, which is based on the speech and image analy-

240   W. Mees, M. Acheroy

sis system ERNEST [14]. The control strategy is implemented as a limited set of instantiation rules, guided by a modified A^* graph-search algorithm.

In [9, 12] we analyze scene analysis domain knowledge and propose to divide it into following categories:

- *scene independent knowledge*: doesn't vary from one scene to another. It can be divided into a general interpretation strategy, object-type specific knowledge and inter-object knowledge.
  - *interpretation strategy*: the order in which different methods are to be applied for a given scene. It is well known that our eyes don't move in a smooth and continuous manner when viewing an image. They go briefly over numerous fixation points separated by jumps and concentrate on those features conveying salient information [6]. The fact of identifying certain objects in the scene gives rise to specific hypotheses and will thus influence the path the eyes follow across the scene.
  A trained image analyst will furthermore steer the focus of his attention along a certain path, based on his training, on a set of Standard Operating Procedures (SOP) and on his experience.
  We will try to imitate the conscious part of the human expert's behaviour using a strategy engine that will direct the order in which different image processing techniques are to be applied.
  - *object-type specific knowledge*: describes which characteristic features allow the system to distinguish a certain object type from other types of objects.
  - *inter-object knowledge*: describes the expected or acceptable geometric configurations for combinations of objects of the same or of a different type.
- *scene related knowledge*: the objects which are present in the scene with their characteristics. This knowledge can sometimes be extracted (partially) a priori from a collateral database, obtained from external information sources or result from previous interpretations of satellite images.

None of the systems mentioned above allows us to represent and integrate all these types of knowledge. A problem solving method which supports the incremental development of solutions, can adapt its strategy to a particular problem situation and allows for different types of knowledge to be accommodated, is the blackboard model [15, 16]. This makes it particularly well suited for complex problems like scene analysis. We therefore propose a system, based on the blackboard model, which can contain all the above defined domain knowledge, distributed over different knowledge sources, where every knowledge source uses its own knowledge representation method.

Based on the blackboard model, a lay-out for our semi-automatic scene analysis system was designed as shown in figure 2.1. The different knowledge sources, interacting solely via the central blackboard, each implement one of the above-described types of knowledge:

Semi-Automatic Analysis of High-Resolution Satellite Images

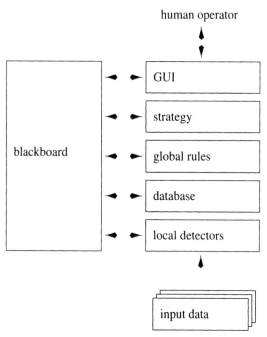

Fig. 2.1. System design.

- *local detector knowledge sources*: the local detectors are scripts which call a series of executables, corresponding to feature extraction and grouping algorithms, with the appropriate parameters and in the required order for detecting a specific type of object in images of a certain spectral band.

  The grouped features correspond with candidate objects which are posted on the blackboard. The local detectors implement the scene independent object-type specific knowledge.
- *global rules knowledge source*: monitors the objects which are posted on the blackboard by the local detectors, selects the relevant rules from a rule-base, applies them to these new objects and posts the resulting instantiations on the blackboard.

  The global rules are expressed as fuzzy production rules [11].

  This knowledge source implements the scene independent inter-object knowledge.
- *strategy knowledge source*: the strategy is represented as a network with nodes and directed arcs between the nodes.

  A node of the strategy network corresponds to a *focus of attention*, defining what the system should be looking for or trying to achieve when this node is activated.

  The links between nodes all have an associated *triggering condition*. When this condition is satisfied for an arc which departs from a previously activated node, the destination node will be activated as well.

This knowledge source represents the scene independent interpretation strategy knowledge.

- *database knowledge source*: a simple geographical database which contains results from previous interpretation operations. One may also load pre-compiled vector maps, for instance obtained from a national mapping agency, into the geographical database as a priori information.
- *graphical user interface (GUI) knowledge source*: allows the user to interact with the interpretation process.

In the actual scene analysis system, the central blackboard consists of two sub-parts:

- the *world-model*, containing the solution hypothesis as it is being constructed. In order to allow for an optimal flow of information when new elements of information (objects and rules) become available, the objects in the world-model are represented, together with the rules which apply to them, as a high-level Petri net graph [10].
- the *blackboard server*, handling all transactions between knowledge sources.

3. Sahara

In order to show image analysts in which way future semi-automatic scene analysis tools may assist them, a research project was initiated, aimed at developing a scene analysis demonstrator. This project was entitled *"Semi-Automatic Help for Aerial Region Analysis"* (SAHARA).

In order to simulate the images as they will be produced by future earth observation satellites, images from airborne sensors were used and their resolution was modified:

- 1.5m resolution visible images, scanned from aerial photographs, taken with a frame camera,
- 4.5m resolution thermal infra-red images, corresponding to the thermal channel of a Daedalus multi-spectral scanner,
- 5m resolution synthetic aperture radar images from the ESAR radar.

As test-case environments a combination of civilian and military airports was chosen. An example of the images for one of the airports is shown in figure 3.1.

The knowledge, as it is incorporated in the system in the form of an interpretation strategy and a set of fuzzy production rules, is the result of discussions between the image processing researchers and the image analysts.

We will now show the content of the world-model after some important phases of the interpretation process for one of the test sites, with one visible, one infra-red and one synthetic aperture radar image available.

For this test we considered a situation where no a priori information from a geographical database or from a human operator is available. The

Semi-Automatic Analysis of High-Resolution Satellite Images 243

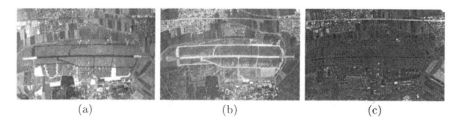

Fig. 3.1. Images: (a) visible image, (b) thermal infra-red image and (c) synthetic aperture radar image.

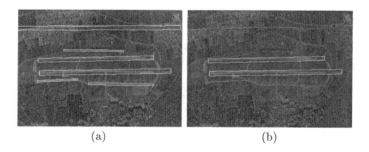

Fig. 3.2. Result images (part 1): (a) detected runway areas in the visible image and (b) filtered runway areas based on geometry.

system then launches the runway area local detector in order to locate the airport in the visible image. The candidate runway areas are filtered based on their geometry and, as shown in figure 3.2, only two parallel runway areas remain, corresponding to the real runway and a full-length parallel taxiway, which could indeed be used as a runway when the main runway should be impracticable.

When two parallel runway areas are found, we know from experience that the airport has been well located by the detector. For this type of airport, the first region of interest on the airport, containing the structure of taxiways and shelters, is generally confined within a relatively restricted region around these two runway areas. This yields the region of interest, shown in figure 3.3.

The taxiway local detector is then launched in the visible and in the infrared image within this search region. The results are also shown in figure 3.3. Nearly the complete taxiway structure has been found, together with a number of false candidates at the top of the image. These false candidates do not result from errors on behalf of the local detectors since they really do fit the local model of a taxiway: a rectangle with certain ranges of possible values for the width and the length. The local detector here suffers from an optical illusion, due to the limited knowledge with which it is equipped.

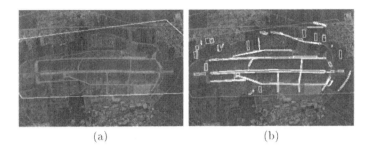

(a)   (b)

Fig. 3.3. Result images (part 2): (a) region of interest based on runway areas and (b) detected taxiways in the visible and infra-red image.

The use of a limited region of interest within which the search is performed, is an important element in the concept of the semi-automatic scene analysis system. Indeed, in this way the search space is gradually reduced. This is a necessary condition for regularizing the otherwise ill-posed image interpretation problem. This gradual reduction is a result of the island-driven approach which was chosen as problem solving method. From a practical viewpoint it is obvious that, if we do not reduce the search region, a large number of false candidates may be generated by the taxiway detectors, especially in a village or in an area with a lot of long and narrow field parcels. Whilst a taxiway is characterized by a well-defined model, the ill-defined shape of a shelter will result in even more false candidates when the search region is not limited to a relatively narrow region around the taxiways.

Because of the poor positioning capabilities of the runway area detector, it is preferred not to use the reported runway areas as "runway objects", but rather to look amongst the candidate taxiways, produced by the taxiway detector which does a much better job at locating the taxiways, for a candidate taxiway which is sufficiently long and wide in order to be a runway. When such a candidate taxiway is found, it is converted into a runway. The result hereof is shown in figure 3.4.

In the meantime, the global rules knowledge source has, every time an object appeared on the blackboard, applied the relevant rules and posted them on the blackboard. As a result, a network was built in the world-model, containing the objects together with the knowledge (in the form of rules) that applies to them. Via this network, objects will strengthen each other's confidence when their relative position fits in with our model of a real world environment. Objects will weaken each other's confidence when they are judged to be conflicting, based on the expert knowledge of the system.

When the confidence values of all objects have stabilized, those objects with a confidence value below a certain threshold are considered to be false candidates and are removed from the world-model. The result is shown in figure 3.4. Most of the false candidates were removed except for one quite

Semi-Automatic Analysis of High-Resolution Satellite Images 245

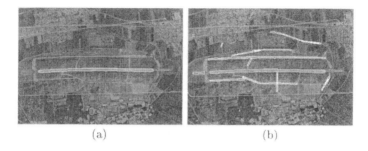

Fig. 3.4. Result images (part 3): (a) finding the runway and (b) taxiways after removing false candidates.

long and one very short candidate taxiway, both at the top of the image. One correct candidate at the bottom-right side of the image was also removed. It should be noted however that both remaining false candidates correspond to a road, which explains why they fit the model of a taxiway rather well. Given a more elaborate model of the taxiway structure on an airport however, they probably would have been eliminated.

Since during this run the system operated in semi-automatic mode, it asked for confirmation by the user between different phases. At this point the user removed the two false candidates, restored the wrongly removed correct candidate and furthermore added the half-circle shaped taxiway at the bottom-left side of the image.

The next phase consists of the detection of shelters. For that purpose, first a search region is determined around the taxiways. Then the shelter detectors are launched within this search region. The search region together with the detection results are shown in figure 3.5.

Fig. 3.5. Result images (part 4): (a) search region for shelters and (b) detected candidate shelters.

4. Conclusions

The demonstrator produced good results for the limited number of test images that were available in the framework of the SAHARA project.

During the development of the demonstrator, the major effort was invested in the development of the system itself. As a result, the knowledge, represented at present in SAHARA, is quite limited. We therefore believe that more complete results can be obtained when a more elaborate strategy and a more complex set of global fuzzy production rules are used.

Acknowledgement. The Sahara project was funded by the Western European Union Satellite Centre (WEUSC) in the framework of its Technical Development Program. The images used in the SAHARA project are the property of the Deutsche Forschungsanstalt für Luft und Raumfart (DLR).

References

1. D. Borghys, C. Perneel, W. Mees, and M. Acheroy, "STANAG 3596 based multispectral information system - feasibility study", Technical Report, Royal Military Academy - Signal & Image Centre, Brussels (Belgium), 1996.
2. R. A. Brooks, "Symbolic reasoning among 3-D models and 2-D images", *Artificial Intelligence Journal*, vol. 17, pp. 285–348, 1981.
3. R. A. Brooks, "Model-based three-dimensional interpretations of two-dimensional images", *IEEE Transactions on Pattern Analysis and Machine Intelligence*, vol. 5, pp. 140–150, March 1983.
4. J. Furnkranz, "The role of qualitative knowledge in machine learning", Technical Report, Austrian Research Institute for Artificial Intelligence, Vienna (Austria), November 1992.
5. O. Grau and R. Tönjes, "Knowledge based modelling of natural scenes", in *European Workshop on Combined Real and Synthetic Image Processing for Broadcast and Video Production*, (Hamburg (Germany)), 23–24 November 1994.
6. K. N. Leibovic, *Science of Vision*. New York: Springer-Verlag, 1991.
7. D. M. McKeown, W. A. Harvey, and J. McDermott, "Rule-based interpretation of aerial imagery", *IEEE Transactions on Pattern Analysis and Machine Intelligence*, vol. 7, pp. 570–585, September 1985.
8. D. M. McKeown, W. A. Harvey, and L. E. Wixson, "Automating knowledge acquisition for aerial image interpretation", *Computer Vision Graphics and Image Processing*, vol. 46, pp. 37–81, 1989.
9. W. Mees, "Scene analysis: fusing image processing and artificial intelligence", *IEEE-Computer Society's Student Newsletter, looking .forward*, vol. 5, pp. 5–8, Spring 1997.
10. W. Mees, "Representing a fuzzy production rule system using a high-level Petri net graph", in *Proceedings of the The Seventh Conference on Information Processing and Management of Uncertainty in Knowledge-Based Systems, IPMU'98*, vol. II, (Paris (France)), pp. 1880–1883, 6–10 July 1998.
11. W. Mees and M. Acheroy, "Automated interpretation of aerial photographs using local and global rules", in *Fuzzy logic and evolutionary programming* (C. H. Dagli, M. Akay, C. L. P. Chen, B. R. Fernandez, and J. Ghosh, eds.),

vol. 5 of *ASME Press series on intelligent engineering systems through artificial neural networks*, (St. Louis, MI (USA)), pp. 459–465, 12–15 November 1995.

12. W. Mees and C. Perneel, "Advances in computer assisted image interpretation", *Informatica - International Journal of Computing and Informatics*, vol. 22, pp. 231–243, May 1998.

13. M. Nagao, T. Matsuyama, and H. Mori, "Structures analysis of complex aerial photographs", in *International Joint Conference on Artificial Intelligence, JCAI-79*, (Tokyo, Japan), pp. 610–616, 1979.

14. H. Niemann, G. Sagerer, S. Schröder, and F. Kummert, "ERNEST: A semantic network system for pattern understanding", *IEEE Transactions on Pattern Analysis and Machine Intelligence*, vol. 12, pp. 883–905, September 1990.

15. H. Penny Nii, "Blackboard systems part I: The blackboard model of problem solving and the evolution of blackboard architectures", *The AI Magazine*, pp. 38–53, summer 1986.

16. H. Penny Nii, "Blackboard systems part II: Blackboard application systems, blackboard systems from a knowledge engineering perspective", *The AI Magazine*, pp. 82–106, summer 1986.

17. A. P. Pentland, "Perceptual organization and the representation of natural form", in *Readings in Computer Vision*, pp. 680–699, Morgan Kaufmann Publishers Inc., 1987.

18. T. Poggio, V. Torre, and C. Koch, "Computational vision and regularization theory", in *Readings in Computer Vision*, pp. 638–643, Morgan Kaufmann Publishers Inc., 1987.

19. H. Reichgelt, *Knowledge Representation: An AI Perspective*. Ablex Publishing Corporation, 1990.

20. I. Tannous, S. Gobert, T. Laurençot, J.-M. Dulac, and O. Goretta, "Design of a multi-sensor system for 3D site model acquisition and exploitation", in *Multi-Sensor Systems and Data Fusion for Telecommunications, Remote Sensing and Radar*, no. CP-595 in AGARD, (Lisbon (Portugal)), pp. 4.1–4.10, The Sensor and Propagation Panel Symposium, 29 September–2 October 1997.

21. A. C. van den Broeck, P. Hoogeboom, and M. van Persie, "Multi-sensor remote sensing for military cartography", in *Multi-Sensor Systems and Data Fusion for Telecommunications, Remote Sensing and Radar*, no. CP-595 in AGARD, (Lisbon (Portugal)), pp. 3.1–3.7, The Sensor and Propagation Panel Symposium, 29 September – 2 October 1997.

22. P. H. Winston, *Artificial Intelligence*. Addison-Wesley, second edition, 1984.

Density-Based Unsupervised Classification for Remote Sensing *

Cees H. M. van Kemenade[1], Han La Poutré[1], and Robert J. Mokken[2]

[1] CWI, Centre for Mathematics and Computer Science,
 P.O. Box 94079, 1090 GB, Amsterdam, the Netherlands.
 e-mail: {Cees.van.Kemenade,Han.La.Poutre}@cwi.nl
[2] Center for Computer Science in Organization and Management (CCSOM),
 Department of Statistics and Methodology, PSCW, University of Amsterdam,
 Sarphatistraat 143, 1018 GD Amsterdam, the Netherlands.
 e-mail: mokken@ccsom.uva.nl

Summary. Most image classification methods are supervised, and use a parametric model of the classes that have to be detected. The models of the different classes are trained by means of a set of training regions that usually have to be marked and classified by a human interpreter. Unsupervised classification methods are data-driven methods that do not use such a set of training samples. Instead these methods look for (repeated) structures in the data.

In this chapter we describe a non-parametric unsupervised classification method, which uses biased sampling to obtain a learning sample with little noise. A density estimation based clustering is then used to find structures in the learning data. The method generated a non-parametric model for each of the classes and uses these models to classify the pixels in the image.

1. Introduction

Classification in remote sensing involves the mapping of the pixels of an image to a (relatively small) set of classes, such that pixels in the same class have similar properties.

Until recently most satellite imagery was at a relative low-resolution, where the width of a single pixel was between 10 metres and 1 kilometre. Currently there are new satellites that produce imagery with high spatial resolution. Such high-resolution images offer new opportunities for Earth Observation applications. In theory, a better classification is possible, as high-resolution images offer more information. In practice, new problems appear. Due to the high spatial resolution, many objects that are too small to locate in low-resolution images are visible in high-resolution images. As a result, the number of different classes that can be detected increases, and the discrimination between classes becomes more difficult. In case of low-resolution imagery one often assumes that a pixel value consists of the spectral vector of the underlying ground cover class plus some Gaussian distributed noise component. It is questionable whether this model is still applicable for high-resolution imagery.

---

* Research supported under US Government European Research Office contract #N68171-95-C-9124.

Apart from the spectral information, there is also spatial information available in the image. This spatial information is used by the human interpreter, but is not used during a purely spectral classification. So, these methods disregard the similarities between neighbouring pixels during classification. Incorporation of the properties of the neighbouring pixels means that we try to exploit spatial structures present in the image.

The outline of our approach is given in section 2. The selection of the learning-samples is discussed in section 3, and the proposed density-based clustering method is given in section 4. Details about the usage of these methods for remotely sensed data are given in section 5. Conclusions and directions for further research follow in section 6.

2. Outline of the Method

In this section unsupervised non-parametric classification is discussed, and density estimation as a method for obtaining this goal is introduced.

In case of a supervised classification method, a human interpreter has to do a pre-classification on part of the image, during which a set of regions is marked and each marked region is classified by the interpreter. Subsequently the supervised method uses these classified regions for obtaining information that will be used to classify the other parts of the image. During unsupervised classification we do not use a set of pre-classified training-samples. The method has to perform a classification of the image based only on the image itself, although some general assumptions about images may be used.

Parametric models assume a certain model for the distribution of a class, where this model has a limited set of parameters. By means of these parameters the model can be fitted to the distribution. Non-parametric models can model almost arbitrary distributions.

We use a classification method that is based on density estimation [4, 5]. The output of the method is a non-parametric clustering of the set of input points. A parametric clustering assumes a certain pre-specified model of the clusters, where an instantiation of this model can be described by a (small) number of parameters. Figure 2.1 gives a schematic representation of the

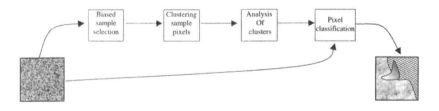

Fig. 2.1. Schematic representation of the classification method.

250 C. H. M. van Kemenade *et al.*

unsupervised classification method introduced in this chapter. On the left of the figure is shown the unclassified image and on the right the classified image. The four steps of the method are:

Sample selection. A biased training-sample is selected from the image. The bias is used to reduce the amount of noise and fraction of mixed pixels in the sample. During this selection spatial information is used. The training sample consists of the spectral vectors of the selected pixels. The sampling method is discussed in section 3.

Clustering. An unsupervised non-parametric clustering method is used to find the clusters. This clustering method is based on density estimation in the sample-space. Details about the clustering method can be found in section 4.

Analysis. The clusters are analyzed. For each cluster the principal component is determined, and the distribution of the pixels when mapped at this principal component is investigated.

Pixel classification. Using cluster information and the original image, a classified image can be generated. The last two steps are discussed briefly in section 5.

3. Biased Sampling of Pixels

A biased sampling method is developed, where the bias is directed towards the selection of pixels containing primarily one ground cover, and with little noise. This is used in the final algorithm inside a stratified random sampling framework.

If we have a region containing a single ground-cover, the noise may be reduced by means of a biased sampling procedure. Here the probability that a pixel is selected, is determined by a similarity measure. We introduce a similarity measure that computes the similarity $sim(p)$ between a pixel p and the pixels in its neighbourhood $N(p)$. We show that the set of pixels with $sim(p) \leq \delta$, i.e. the pixels that are relatively similar to their neighbours, comprise a biased sample of pixels containing relatively little noise. Let the function $d(s_1, s_2)$ denote the spectral distance between two pixels. A similarity measure can be obtained by calculating $d(\cdot, \cdot)$ for all neighbours. Furthermore the rank with respect to their distances is given by $r(i)$. The similarity measure used is given by:

$$sim(p) = \sum_{i \in N(p)} W_{r(i)} d(s_p, s_i)$$

where p is the current pixel, $N(p)$ is its neighbourhood, and $\mathbf{W}$ is a weight vector such that $|\mathbf{W}| = 1$. The best choice of $\mathbf{W}$ will depend on the structures one is interested in. If thin structures should be detectable, such as rivers or

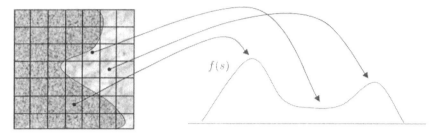

Fig. 3.1. A square region containing two types of ground cover and the corresponding density function obtained when mapping all pixels to the spectral space.

roads that have a width comparable to the width of a pixels, then W_1 and W_2 should be the only non-zero values. If one is not interested in such thin structures, it is better to have $W_k > 0$ for higher values of k, as this puts a stronger emphasis on the similarity of a pixel with respect to its neighbours, and therefore a stronger emphasis on the homogeneity of the pixel. The similarity measure $sim(p)$ is likely to have a low value if p belongs to the same class as its neighbours and contains relatively little noise. A theoretical analysis is given in [2]. There it is shown that if we have a set of t pixels drawn according to a Gaussian distribution, and the biased sampling procedure described above is used, then the corresponding sample is concentrated around the center of this distribution.

In the previous paragraph we introduced a biased sampling method, and we have shown that this biased sampling method is able to select a set of relatively noiseless pixels in case of a one-class region. Now we will show intuitively that the sampling procedure also works with multiple classes. Figure 3.1 shows a region containing two classes, i.e. two types of ground cover. The figure also shows the density function obtained when mapping all pixels from the square patch to the spectral space. The arrows show the locations of three different pixels in the density-graph. On the left of the graph we see a peak that corresponds to the main ground cover within the square patch. On the right we see a lower peak that corresponds to the other ground cover. In between those two peaks, we see a ridge that corresponds to the mixed pixels that are located on the boundary of the two sub-regions. If we compute the similarity measure $sim(p)$ for these three pixels, then it is likely that the two non-mixed pixels that are located close to a peak have a lower value of $sim(p)$ than the mixed pixel. Therefore the two pixels containing a single ground cover are more likely to be selected than the mixed pixel.

To summarize, the local spatial homogeneity is exploited by means of the assumption that pixels in the spatial neighbourhood $N(p)$ of point p are likely to belong to the same class as point p. Pixels with lots of noise and mixed pixels are both likely to have a large value of $sim(p)$. So, if we select a biased sample for which $sim(p) \leq \delta$, then this sample is likely to contain the pixels with little noise and containing only a single ground-cover.

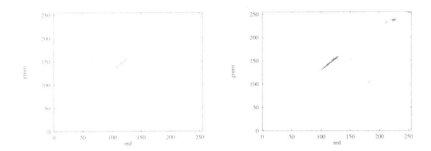

Fig. 3.2. Results for a real image with stratified random sampling (left) and biased sampling with $k = 3$ (right).

The biased sampling method was applied to some real images, and was compared to a stratified random sample. All images are three-band images, with a sample size of 4000 pixels. The stratified random sample is taken by dividing the image in a set of non-overlapping rectangular areas, and taking a random pixel from each area. The distribution of the spectral vectors of such a random sample is compared to the distribution when using a biased sampling method. During the biased sampling we used $k = 3$. The value of δ was adapted such that exactly 4000 pixels where selected, so in fact we select the 4000 most homogeneous pixels. The results are shown in figure 3.2. We observe that the biased sample is revealing more structure than the random sample. This shows, that the biased sampling method is able to highlight the cluster structure in the image-data.

In many images, some of the important ground cover types cover only a tiny fraction of the image. For example, a road that passes through an image of 1000 × 1000 pixels will only cover approximately 0.1% of the image. The human eye can easily detect such a road, as it is very sensitive to local spatial structure in an image. When taking a random sample of this image, and mapping the points to the spectral space, all this spatial information is lost, and it is unlikely that we are able to detect a cluster that is covered by such a small number of pixels. We have also developed an extended sampling method that biases the sample to points belonging to clusters that cover only small part of the image. This method uses density estimation to compare the local density, in a small patch of pixels, to the global density of a pixel. For a complete description of this sampling method, see [2].

4. Hierarchical Clustering

In this section we give a detailed description of the hierarchical clustering method we have developed. This method takes a set of points as an input and produces a set of clusters. Simultaneously it computes a measure for the

Fig. 4.1. Density based hierarchical clustering.

separability of all the clusters. First a description of a density estimation method is given, that forms the basis for the clustering algorithm. Furthermore in order to sketch the basic approach of the clustering method we start with a rather intuitive example.

In the clustering method, a density estimation procedure [4, 5] is used. Let a data-set consist of points drawn according to an unknown distribution. Now a density estimation method constructs a density function based on this data-set, which explains the distribution of the data-points. This density function is related to a distribution function, which should approximate the original, but unknown, distribution function that generated the data.

We use a density estimation method that is based on the distance to the kth nearest neighbour, the k-NN estimator [5]. To compute the kth nearest neighbour, we use an advanced data structure, viz., the optimized kd-trees [1]. We use these trees because a brute-force computation of the nearest neighbours in a data set of size N will require $N-1$ distance computations, while by means of kd-trees the same computation will only take a number of steps proportional to $\log N$. Given the distance to the kth nearest neighbour $d_k(x, \{x_i\})$, the local density is computed by the formula

$$\hat{f}(x) = \frac{k}{d_k(x, \{x_i\})^d N V_d(1)}$$

where N is the sample size and $V_d(1)$ is the volume of a d-dimensional sphere with radius one.

The operation of the hierarchical clustering method is represented graphically in figure 4.1. The left side of this figure shows a density curve over a one dimensional space. When observing a single curve we see that the sample space contains three locations with a density that is higher than the density in the neighbourhood. The left one is the densest of these three. The horizontal line in figure 4.1 represents the decreasing threshold, used by the method. On the left of the figure the threshold is still high. Each connected region above the threshold results in two separate clusters, so in this case we have two clusters, denoted by the solid line-segments below the graph. On the right side of figure 4.1, the threshold is lowered. The density region in between the two clusters is above the threshold. As a result the two clusters have been merged into a single cluster. If we lower the threshold even further the points corresponding to the third peak will be detected. In practice the

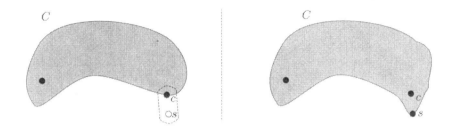

Fig. 4.2. Extending a cluster by one point.

density function of a data-set is not known, but it may be approximated by means of density estimation methods.

The algorithm keeps track of three lists. The first list contains all datapoints, the second list contains all clusters, and the third list contains delayed cluster merge operations. Initially the last two lists are empty. For each of the points the local density is estimated, and the list of points is ordered on decreasing density. So a position closer to the start of this list corresponds to a higher density estimate. If the data is distributed according to the density given in figure 4.1, then the sample-points close to center of the left cluster will be at the start of this list. If we process a sample-point s, then we set the current threshold equal to its local density estimate. Before actually processing the point, we first process all delayed merge operations that have to be performed at the current threshold. Once an operation is performed, it is removed from the merge operation list. Now we generate a new cluster containing sample-point s only, and we add this new cluster to the list of clusters. Next we have to determine when this new cluster can be merged with any of the other clusters. The computation for a single cluster C is shown in figure 4.2. The new cluster consists of a single point denoted by the small circle. The cluster C is denoted by the grey region. The density at which this merge operation can be performed is computed as follows. The nearest neighbour of s in C is located, let us denote this point by c. The joint density of s and c is taken as an estimate for the density at which the cluster containing s is merged with C. This density is computed by constructing the cylindrical envelope of the k-NN neighbourhoods of points s and c. The cylindrical envelope is a cylinder with rounded sides. Given the volume of the cylindrical envelope and the fact that it contains at least $2k$ points, we can compute the density over this volume. Using this approach, the separability of two clusters is determined by the density of the densest connection between the two clusters. Also, the algorithm will be sensitive to the minimal allowed density that we use during the clustering process. If this density is too low, all clusters are likely to be merged in a single cluster, if this density is two high, then low density clusters, corresponding to classes of ground cover that are relatively seldom in the image, will not be detected. If the dimension of the

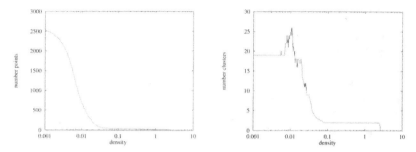

Fig. 4.3. The number of points classified (left) and the number of clusters (right) as a function of the density

spectral space increases, its sensitivity with respect to the minimal density will increase. Therefore a second measure for the separability of two clusters is introduced, and is used to put an additional restriction on the merging of two clusters. The new measure is the separation measure, given by the formula

$$\frac{v}{\min\{p_k, p_l\}},$$

where p_i is the maximum of the density in cluster i, and v is the density in between the two clusters. This measure determines the density ratio between the peak density of the low-density cluster, and the density in between the clusters. The value of the separation measure is between zero and one. A value close to one means that the low-density cluster has a density that is close to the density of the region in between the clusters. As the densities we are working with, are estimated densities, we can merge the two clusters in this case. If the value is close to zero, then the low-density peak has a significant higher density than the region in between the two clusters, and it is likely that the low-density cluster corresponds to another class. Thus, no merge is performed. We also state that the theoretical estimated time complexity can be quantified as $\leq d \cdot (NC\log N)$ steps for some constant d, where N is the number of data-points and C is the maximal number of clusters that exist simultaneously.

When the hierarchical Clustering method is applied to a data-set, it produces an output which is shown in figure 4.3. The left graph shows the number of points that have been classified as a function of the density. The right graph shows the actual number of clusters obtained as a function of the density. If we choose a new density threshold that is larger than the current *minimal Density*, then the corresponding classes can be generated almost instantly, based on an internal data-structure generated by the hierarchical Clustering method. This, in contrast to most other clustering methods which require that one determines the number of clusters beforehand, and a change in the number of clusters requires a new run of the method.

5. A Remote Sensing Application

The hierarchical clustering method produces a set of classes. A pixel-based classification method is needed to assign all pixels to a class. The clustering method can produce classes that correspond to non-convex regions in the spectral space. Therefore, the pixel-based classification method should be non-parametric. We use a nearest-neighbour classifier. A pixel is classified by finding the (spectral) nearest pixel in the learning-sample, and assigning the class of this pixel. We use a kd-tree to find the nearest neighbour of a pixel in a number of steps proportional to $\log N$, where N is the size of the learning sample.

We have applied the methods presented in this chapter to remotely sensed data. We have tested the method on a high-resolution three-band aerial photograph of 500 × 538 pixels and on 7-band Landsat scene with 960 × 1130 pixels. In the case of the Landsat images the sixth band, which corresponds to thermal emission, was removed from the data-set, following [3]. In both cases we used a sample of 4000 points for learning. The tests were performed on a SUN workstation running at 180 MHz. The sampling method used during this experiment involves the computation of local and global statistics, as described in [2]. For the three-band image the generation of the biased sample took approximately 33 seconds, and a classification containing 13 clusters was obtained in approximately 16 seconds. For the six-band Landsat image the generation of the biased sample took approximately 95 seconds, and a classification containing 26 clusters was obtained in approximately 59 seconds. When using a sample containing 11,000 points for the 7-band image the sampling step takes 460 seconds, and the clustering step takes 421 seconds.

In case of the 7-band Landsat scene we also had a map, showing the results of a supervised classification of the land usage of part of this region. The resolution of the map and the Landsat scene were different, and geometric corrections were applied to the map. Therefore, we can only give a qualitative comparison between the map and the classification obtained by our method. The types of ground-usage shown in the map are agriculture, industry, city, residential, water, and natural vegetation. In comparison to the land use map our method finds more classes. For example, we detect many different classes in the agriculture region. It is interesting to see that most of the regions found are rectangular regions, that are aligned with the neighbouring regions. The shape, orientation, and size of these regions correspond to the typical plots of land in agriculture regions. It seems that our method is able to discriminate between the different types of agricultural use of the land in this region. Also the water regions, that cover only approximately 0.6% of the total surface, come out clearly. There also detected two classes for the urban regions, the first class is mainly located near the center of urban regions, while the second class is located more towards the boundaries of the urban areas. This may correspond to the discrimination between city and residential area in the map. The boundaries between city and·residential are different in our case, though we can easily imagine that these boundaries are not very well defined, and we see it as a promising result that our method already detects different types of urban areas. There is only little industrial activity in this region, and it seems like the industrial regions are classified as residential area's in our method. The areas with natural vegetation are split into two classes.

When doing the same analysis for a lower value of the separability parameter we get a more course grain classification. The agriculture regions are less diverse, and the city and residential area are merged in a single class. Natural vegetation is covered by a single class too.

6. Conclusions

We developed a biased sampling method and a hierarchical clustering method. The sampling method exploits spatial information in order to select those pixels that correspond to a single ground cover, and contain relatively little noise. The sampling method has been tested by means of a theoretical model and on real data. In both cases we observe that the clusters in the data-set are more clearly present, when using a biased sample instead of a random sample.

The hierarchical clustering method is a fast, unsupervised clustering method that takes a set of points as an input and produces a set of non-parametric classes describing the input-data. The method is purely data-driven, and therefore the number of clusters obtained is dependent upon this data-set. Apart from the sample-sizes and neighbourhood sizes, the method uses a separability parameter. This parameter determines under what conditions two clusters can be merged into a single cluster, and therefore affects the final number of clusters. Furthermore this parameter has an intuitive basis in terms of the ratio of the peak-densities of clusters and the density of the ridge connecting the clusters.

Future work includes the development of non-parametric models from the learning-data by means of radial basis neural networks, the use of evolutionary computation methods to search for models that allow a demixing of clusters consisting of multiple classes, and the usage of a Bayesian approach to exploit the spatial structures by computing prior probabilities over a spatial neighbourhood.

References

1. J. H. Friedman, J. L. Bentley, and R. A. Finkel, "An algorithm for finding best matches in logarithmic expected time", *ACM Transactions on Mathematical Software*, vol. 3, no. 3, pp. 209–226, 1977.

2. C. H. M. van Kemenade, J. A. La Poutré, and R. J. Mokken, "Density-based unsupervised classification for remote sensing", Technical Report SEN-R9810 (also available via http://www.cwi.nl/~hlp/PAPERS/RS/dense98.ps), CWI, Amsterdam, the Netherlands, 1998.
3. J. A. Richards, *Remote sensing digital image analysis*, Springer-Verlag, Berlin, 1993.
4. D. W. Scott, *Multivariate Density Estimation*, Wiley series in probability and Mathematical Statistics, John Wiley & Sons, INC., New York, 1993.
5. B. W. Silverman, *Density estimation for statistics and data analysis*, Monographs on Statistics and Applied Probability, Chapman and Hall, London, 1986.

Classification of Compressed Multispectral Data

Frank Tintrup, Cristina Perra, and Gianni Vernazza

Department of Electrical and Electronic Engineering University of Cagliari,
Piazza d'Armi, 09123 - Cagliari, Italy.
e-mail: {tintrup,cris,vernazza}@diee.unica.it

Summary. The requirements of increasing resolution in remote sensing are leading to excessively large data volumes. Powerful compression techniques are therefore needed to reduce bandwidth needed for satellite downlinks and to ease storage requirements. In this chapter, we report on a comparison of the principal lossy compression algorithms. The analyzed techniques are Vector Quantization (VQ), JPEG and Wavelets (WV), which have been applied to multispectral remotely sensed images. Prior to compression by the different algorithms the Karhunen Loeve Transform has been used to remove the interband correlation and produce the principal components of the image. A supervised classification with the well known K-Nearest Neighbour algorithm for remote sensing applications has been used for quality evaluation while the Mean Square Error (MSE) was calculated in the second order.

1. Introduction

Multispectral remotely-sensed (RS) images contain enormous data volumes. They therefore pose significant storage problems (especially at high resolution) and require high bandwidth for downlinks. Moreover, in viewing, analyzing, and classifying such images with many spectral bands, efficient methods are required to reduce the redundant information. Data compression could therefore play an important role in analysis and classification of remotely sensed imagery.

Applications of RS imaging include change detection (in which images of the same ground area may be acquired and stored for long periods), Earth resources monitoring and terrain classification. For most of these applications, multispectral images must be acquired by satellite or aircraft mounted sensors. Automated or semi-automated image processing tools are then used to identify and classify different land cover types such as agricultural areas, urban areas, forests etc. Landsat Thematic Mapper (TM) images, which might be used in such applications for example, usually occupy about 150 Mbytes. After acquisition they are generally stored or transmitted to ground stations without using any compression tools, thus requiring a very large bandwidth.

Some interesting preliminary results in the field of lossless compression techniques have been published [13], some of which include extending the JPEG standard to RS images [4, 9]. Good results have also been obtained with advanced Vector Quantization (VQ) techniques. There are still no results available in the literature, however, concerning the automatic classification of compressed multispectral remotely sensed images.

2. Methods

2.1 Karhunen Loeve Transform

Usually, remotely sensed multispectral images contain a large amount of interband correlation due to the co-located sensors and the spectral overlap of the bands. The best known technique to exploit this correlation accurately is the Karhunen Loeve Transform (KLT) which produces the principal components of the image. The KLT, an orthogonal transformation, provides the minimum mean square error (MSE) during decorrelation by discarding the high index coefficients in the transformed space and maximizing the energy in the fewest number of coefficients. The highest energy is concentrated in the transformed space corresponding to the largest eigenvalues.

2.2 JPEG Algorithm

This chapter deals only with single-component (grayscale) images. Therefore only the DCT based mode of operation is considered, essentially the compression of a stream of 8×8 pixel blocks. For detailed information on this algorithm see the paper by Wallace [12].

2.3 Vector Quantization

A memoryless vector quantizer (VQ) consists of an encoder which assigns to each input vector a channel symbol from a specified channel symbol set, called the codebook, and a decoder which assigns to each channel symbol of the codebook an output value in a reproduction alphabet [6]. The generation of the codebook is done by the well known LBG algorithm [8].

In this approach to RS image coding the VQ operates on small, 2-dimensional rectangular block samples of 2×2, 3×3 or 4×4 image pixels. The decoded image quality is mainly dependent on the block and codebook size.

Typical characteristics of decoded images are, in particular, the poorly reproduced edges and the known "blocky" effect due to codeword edges. Some particular solutions to this, e.g. the construction of segmented codes and separated codebooks for edge and texture information, have been studied [5].

2.4 Wavelet Transform

More recently the wavelet transform has been used for image data compression. This technique is based on adaptive vector quantization of wavelet coefficients. This promises high compression rates with good image quality. It usually performs better than JPEG and VQ, both in quantitative (MSE) and qualitative terms, and is free of block distortion found with VQ and JPEG [1].

Several methods have been presented in the literature for wavelet-based image compression. A certain number of approaches propose vector quantization of the wavelet transformed coefficients in different sub-bands [2, 14].

The wavelet representation of an image is composed of the approximation of the signal at low resolution plus a set of details at several resolutions. The image at low resolution is a low-pass version of the original one, while the details contain the information at high frequencies. The signal of each sub-band is found through an iterative algorithm which decomposes the original signal into four more detailed ones where each signal contains the information regarding a particular frequency band and orientation, see figure 2.1.

The reconstruction algorithm is strongly related to the decomposition technique in that the complete signal is found again through a pyramidal algorithm, taking into account the low-pass signal and the set of details.

The wavelet representation used was studied by Lemarie and Battle [3, 7] and corresponds to a multiresolution approximation constructed from cubic splines. Particular attention has to be paid to the encoding of the low-pass version (LL sub-band) as introduced errors could propagate in the reconstruction phase, resulting in worse image quality. The encoding is therefore done by the lossless bidimensional DPCM technique defined in [10] which takes advantage of the correlation between adjacent pixels in all the directions (horizontal, vertical and diagonal). This technique involves initially a 1-D DPCM applied to the first row and first column of the image. Then each pixel value is predicted with a linear combination of its three nearest pixels values where the prediction error is coded.

The statistical distribution of the wavelet coefficients at a fixed resolution and orientation is a symmetric distribution with a nearly zero mean and

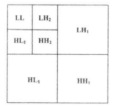 

Fig. 2.1. Wavelet representation of first KLT sample.

small variance. It is often modelled as a Laplacian distribution even though it falls off more rapidly and is therefore better approximated by the Generalized Gaussian Distribution [11, 15]. The properties of the nearly Gaussian distribution allow a distinction between certain zones in each sub-band, characterized by greater energetic and informative contents, called the "active zones" where an accurate coding process improves the reconstructed image quality. An extensive empirical analysis of different image types has lead to a heuristic algorithm to identify the active zones in which a threshold process is applied to the histogram of the quantized wavelet coefficients histogram of each sub-band.

Once the process is terminated, a binary-value "mask" of each sub-band is extracted which contains information regarding the position and shape of the active zones. The coefficients belonging to the non-active zones are not considered any more due to their very low informative content, and they constitute the background of the sub-band. Each mask is then re-scaled and logically summed to the one obtained at the lower adjacent resolution and the same orientation. Through this process we obtain three masks where each contains the information regarding the active zones of the sub-bands at the same orientation (LH, HL and HH). Finally, the masks are logically summed to obtain a unique mask which is encoded by optimized run length coding.

The active zone coefficients contain the main part of the energy and information of the relative sub-band and have to be therefore encoded accurately. We used an adaptive vector quantization where parameters have been chosen depending on energy considerations. Generally, the HH sub-bands contain less information than LH and HL. Also a sub-band at low resolution contains more energy than sub-bands at the same orientation and higher resolutions -a reason for a more accurate quantization of the sub-bands at lower resolution. Moreover, the dimensions of codebooks and code-vectors were chosen accordingly to the variation of the MSE during the LBG algorithm [11]. In our experiments, we generated specific codebooks for each sub-band at each resolution and orientation from active zones of sub-bands of the images belonging to the training set.

2.5 K-NN Classification

To evaluate the quality of the compression algorithms we used supervised classification by the well known K-NN (K-Nearest Neighbor) non-parametric technique. The K-nearest-neighbor rule can be expressed as follows: Classify the unknown sample by assigning it the label most frequently represented by the K nearest samples measured by the Euclidean distance. The value of the parameter K (21 in our case) was calculated by the square root of the number of features in the training set, an empirical rule for this classifier. This rule does not describe the optimal value of K but empirically it is known that it will be approximately the optimal value.

3. Experiments

The test set used to compare the discussed compression algorithms consisted of the six visible mid infrared bands of an image of an agriculture area in the UK, acquired by a TM sensor. Portions of the original image data consisting of blocks of 250×350 pixels were used with 8 bit quantization per pixel value.

The six image samples were spectrally decorrelated by the KLT which produces the principal components. Then, each of the six principal components were compressed by the lossy JPEG algorithm, varying the compression rates. The standard quantization tables were used which were empirically defined by JPEG and are appropriate for most applications. Due to the application of the KLT before compression, most quantized DCT coefficients of the less significant image planes have a zero result and improve therefore the run length coding of the JPEG algorithm. This therefore yields higher compression ratios for these image planes, in comparison with the more significant planes, without decreasing their quality.

Thereafter, the VQ was applied to the principal components. For each of the decorrelated images a separate codebook with 256 vectors from images of the same type and with similar characteristics was computed. For different compression ratios, the block size was from 2×2 to 5×5 pixels.

Finally, the novel adaptive compression of the wavelet coefficients was applied where the redundant information in the decorrelated image planes was reduced by coding the wavelet coefficients. The coefficient-histogram of each sub-band from the image planes was analyzed and only the active zones were then quantized. The sub-band LL was losslessly DPCM-coded while the wavelet coefficients of all other sub-bands where coded by an adaptive VQ.

Finally, the inverse KLT was applied to all the decompressed image planes to reconstruct the image portions -part of the initial remotely sensed images. The reconstructed images were then fed to a K-NN classifier where a supervised classification was performed. This was compared to the results obtained with the original (uncompressed) images using a second order Mean Square Error (MSE), which is a visually interesting measure. In figure 3.1 we show a simplified and schematic overview of the work carried out.

4. Results

The results of the evaluation measures "Decrease of Correct Classification" (DCC) and MSE are shown in figures 4.1 and 4.2 respectively. The decoded images are shown in figure 4.3. The Compression Ratios (CR) of the JPEG and Wavelet algorithm are average values of the six single image planes as these algorithms compress the less significant KLT transformed image planes much better than the others. Also the presented percentages DCC are average values, compared to the average values from "Correct Classification" (CC) of the five known single classes, taking into account the amount of training and

264 F. Tintrup et al.

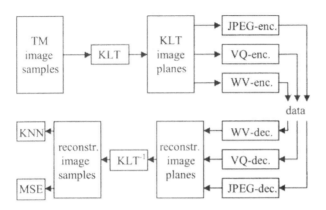

Fig. 3.1. Schematic overview of the work carried out.

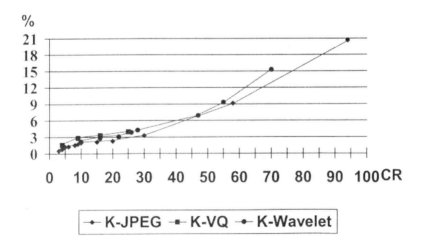

Fig. 4.1. Graphical representation of "Decrease of Correct Classification".

Classification of Compressed Multispectral Data 265

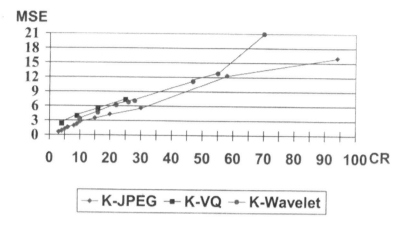

Fig. 4.2. Graphical representation of MSE.

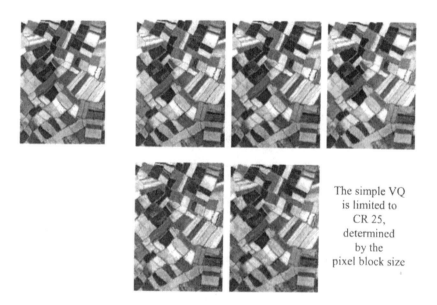

Fig. 4.3. Original and reconstructed TM image planes, upper images, from left, original most significant image band 7, KLT-JPEG, KLT-Wavelet and KLT-VQ at compression rates 25, lower images at compression rates 50, from left KLT-JPEG, KLT-Wavelet.

test samples. For our comparison of the compression algorithms the detailed results of the single classes are not of high interest. Moreover, we verified that the application of our KLT algorithm itself did not manipulate the results significantly.

Note that the decoded "wavelet" images, shown in figure 4.3 have a somewhat better visual quality compared to the images from KLT-JPEG and KLT-VQ at the same compression rates, even though they are classified less accurately by KNN and have a higher MSE. This is because of the very low block effect due to the DPCM compression of the sub-band LL. Instead in these images some kind of "unsharpness" or blurring is notable in certain areas. Note, in particular, that these areas contain not very significant information which has been therefore lost in part during coding of just the active image zones. We see that the KLT-JPEG algorithm yielded the lowest DCC and MSE values compared to the other techniques for equivalent compression ratios, while the worst results were obtained with the KLT-VQ algorithm. Generally, the K-NN classifier was less sensitive than the MSE with all the algorithms used. This was evident also by the decrease in visual quality with even a quite low DCC.

5. Conclusions

From the techniques tested, KLT-JPEG was found to best fulfill the requirement for compressing multispectral TM images i.e. that of minimizing DCC by K-NN. Moreover, the other two algorithms both require an updating of their codebooks depending on the characteristics of the input image samples. The results obtained regarding the KLT-Wavelet approach were surprising, as this approach is usually more efficient in comparison with the others for many applications. The main reason for the unexpected low efficiency of this recent technique, is that the algorithm is not well optimized for automatic classification of remotely sensed images, even though it performs well in browsing applications where good visual quality is the main objective [1]. Usually in these kind of applications the WV algorithm finds many active zones which can be coded to provide good visual quality even at very high compression rates.

To take advantage of the capacity of WV coding for these applications we propose an adaptive quantization and coding of the KL- and then Wavelet-transformed image samples. This optimization may depend on the characteristics of the remotely sensed images. The main possibility of obtaining this optimization with the proposed Wavelet algorithm is to vary the quantization factor in the pre-coding phase and then to additionally adapt the block size for the VQ depending on the analyzed image plane.

References

1. M. Ancis and D. D. Giusto, "Image data compression by adaptive vector quantization of classified wavelet coefficients", in *Proceedings of the IEEE Pacific Rim Conference on Communication, Computers and Signal Processing, PACRIM'97*, Victoria, Canada, pp. 330–333, 1997.
2. M. Antonini, M. Barlaud, I. Daubechies and P. Mathieu, "Image coding using vector quantization in the wavelet transform domain", in *Proceedings of the IEEE International Conference on Acoustics, Speech and Signal Processing*, Albuquerque, NM, pp. 2297–2300, 1990.
3. G. Battle, "A block spin construction of ondelettes - Part I: Lemariè functions", *Communications on Mathematics and Physics*, vol. 110, pp. 601–615, 1987.
4. M. Datcu, G. Schwarz, K. Schmidt and C. Reck, "Quality evaluation of compressed optical SAR images - JPEG vs. wavelets", in *Proceedings of the International Geoscience and Remote Sensing Symposium, IGARSS'95*, Florence, Italy, vol. III, pp. 1687–1689, 1995.
5. A. Gersho and B. Ramamurthi, "Image coding using vector quantization", in *Proceedings of the IEEE International Conference on Acoustics, Speech and Signal Processing*, vol. 1, pp. 428–431, 1982.
6. R. M. Gray, "Vector Quantization", IEEE press, New York, USA, 1990.
7. P.G. Lemariè, "Ondelettes à localisation exponentielles", *Journal of pure and applied Mathematics*, vol. 67, pp. 227–236, 1988.
8. Y. Linde, A. Buzo and R. M. Gray, "An algorithm for vector quantizer design", *IEEE Transactions on Communications*, vol. COM-28, pp. 84–95, 1980.
9. A. J. Maeder, "Lossless JPEG compression of remote sensing imagery", in *Proceedings SPIE - The International Society for Optical Engineering*, vol. SPIE 2606, pp. 12-2-132, 1995.
10. A. N. Netravali and B. G. Haskell, "Digital Picture - Representation, Compression and Standards", Plenum Press, 1995.
11. M. Vetterli, J. Kovacevic, "Wavelets and subband coding", Prentice Hall, USA, 1995.
12. G. K. Wallace, "The JPEG Still Picture Compression Standard", *IEEE Transactions on Consumer Electronics*, vol. 38, no. 1, pp. xviii-xxxiv, 1992.
13. J. Wang and K. Zhang, "An algorithm for constrained optimal band ordering of multispectral remote sensing images in lossless data compression", in *Proceedings of the International Geoscience and Remote Sensing Symposium, IGARSS'95*, Florence, Italy, vol. III, pp. 1684–1686, 1995.
14. P. H. Westerink, J. Biemond, D. E. Boekee and J. W. Woods, "Subband coding of images using vector quantization", *IEEE Transactions on Communications*, vol. COM-36, no. 6, pp. 713–719, 1988.
15. P. H. Westerink, "Subband coding of images", Ph.D. Thesis, Delft University of Technology, 1989.

Part V

Segmentation and Feature Extraction

Detection of Urban Features Using Morphological Based Segmentation and Very High Resolution Remotely Sensed Data

Martino Pesaresi and Ioannis Kanellopoulos

Space Applications Institute, Joint Research Centre,
European Commission, 21020 Ispra, Varese, Italy.
e-mail: {martino.pesaresi,ioannis.kanellopoulos}@jrc.it

Summary. The spatial resolution of remotely sensed imagery has improved considerably during the last few years and will increase dramatically in the near future due to the imminent launch of the new generation very high-resolution sensors. For urban applications in particular, with the spatial properties of the new sensors it will be possible to recognise, not only a generic texture window with specific urban characteristics, but also to detect in detail the objects that constitute the "urban theme". However, the improvement in the spatial resolution may result in a decrease of the accuracy of automatic classification techniques, if only the standard multi-spectral analysis procedures are applied. In this chapter a per-segment segmentation procedure is presented, based on the gray-scale geodesic morphological transformation and has been successfully utilised to detect built-up objects using only the 5 m spatial resolution panchromatic data of the IRS1-C satellite. The imagery is subsequently classified on a per-segment basis using a multi-layer perceptron neural network classifier.

1. Introduction

The spatial resolution of images offered by Earth Observation satellites until recently was not sufficient to provide detailed topographic features such as shape and structure information. This is particularly important in applications such as topographic mapping and monitoring of the urban environment. New generation sensors however, already provide or will provide very shortly spatial resolutions of 1 to 5 metres, which will change considerably the spatial quality of the remotely sensed imagery. This, on the other hand, may result in a greater ambiguity in the definition of land cover/use classes, when compared to lower resolution imagery, and the accuracy of the automatic recognition of urban areas will be decreased if only the standard multi-spectral analysis procedures are applied. Therefore it is apparent that new analysis tools are needed, to exploit shape and structural information inherent in very high spatial resolution imagery. Some of these tools have already been applied to other fields, for example, aerial photography or medical imaging.

In this chapter we exploit the use of morphological operators with very high resolution remotely sensed imagery to detect man-made objects within urban areas. In particular a segmentation procedure is presented based on gray-scale geodesic morphological transformations using 5 m spatial resolution panchromatic data from the IRS-1C satellite.

Initially the IRS-1C panchromatic (PAN) imagery is pre-processed using a set of morphological transformations, namely, successive opening and closing operators for morphological feature extraction. The output of this stage consists of a set of structural raster layers, which provide the morphological characteristics used in the segmentation phase. For each structural layer and all resulting regions a set of statistics is computed which includes the mean and standard deviation and a set of simple shape descriptors. Finally a multi-layer perceptron neural network is first trained on the region statistics and subsequently classifies the imagery into seven man-made ground classes.

2. Mathematical Morphology

Mathematical Morphology is an approach to image processing that deals with the geometric structure of objects within an image. It consists of a set of operators that are particularly suited for the extraction of shape or structural information of an image. These operators were first examined systematically by Matheron [10] and Serra [17, 18, 19], and are an extension of Minkowsky's set theory. A morphological operator is governed by a small *mask*, M, called *structuring element*. When applied to an image, the operator returns a quantitative measure of the geometrical structure or shape contents of the image, in terms of the structuring element. The two fundamental morphological operations for binary images are *dilation* and *erosion* defined as [5, 6],

$$\delta_M(X) = \{p : M_p \cap X \neq \emptyset\} \quad (2.1)$$

and

$$\epsilon_M(X) = \{p : M_p \subseteq X\} \quad (2.2)$$

respectively. Equation 2.1 states that the dilation of a set of pixels X by the set of pixels or structuring element, M, is the set of all pixels p for which the intersection between X and M_p is not an empty set. Equation 2.2 expresses the erosion of the set of of pixels X by the structuring element M as the set of all pixels p such that M_p is completely contained in X.

The *binary opening* operator is defined as the erosion of an object followed by dilation using the same structuring element:

$$\gamma_M = \delta_M(\epsilon_M(X)) \quad (2.3)$$

Similarly the *binary closing* operator is defined as:

$$\phi_M = \epsilon_M(\delta_M(X)) \quad (2.4)$$

The concepts of the binary operators described above may be extended to gray-level images. The main difference between binary and gray-level operators is that the transformations are performed in three dimensions (x, y)

coordinates and gray-level value) as opposed to two dimensions for the binary operators. Let f be a gray-level function and M is a "flat" structuring element (i.e. all values set equal to zero). Then the dilation and erosion of f by M are defined as:

$$\delta_M(f(x)) = \{max(f(u)) \mid u \in M_x\} \quad (2.5)$$

and

$$\epsilon_M(f(x)) = \{min(f(u)) \mid u \in M_x\} \quad (2.6)$$

where $f(u)$ is the value of the function f at point u.

Morphological operations can use either the classical Euclidean metric or non-Euclidean or *geodesic metric*, i.e. the "reconstruction" morphological filtering. Filters by reconstruction form an important class of morphological connected filters [3, 20], that have proven to be very useful for image processing [16]. The reason is that they do not introduce discontinuities and, therefore, they preserve the shapes observed in an input image. That is, there is a lower shape noise introduced by the interaction between the shape of the structuring element and the shape of the image to be transformed. The *geodesic distance* metric, was introduced by Lantuéjoul and Beucher [8] in the framework of image analysis. It is defined as follows: given a set X the geodesic distance between two pixels a and b is the length of the shortest path joining a and b which are included within X. The notion of geodetic distance is illustrated in figure 2.1. Figure 2.2 illustrates the operation of the geodesic dilation of set B within A using structuring element M. In mathematical terms this is defined as,

$$\delta_M^A(B) = (\delta_M(B) \cap A) \cup B \quad (2.7)$$

So far mathematical morphology has not been widely applied in remote sensing applications and only a few authors have applied morphological operators in the analysis of remotely sensed imagery (see, for example,

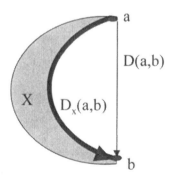

Fig. 2.1. Geodesic distance $D_x(A, B)$ between two points a and b within object X.

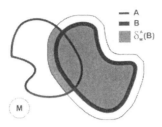

Fig. 2.2. Illustration of the geodesic dilation of Y within X.

[4, 9, 11, 15, 22]). Instead the main applications areas have been medical imaging, material sciences and robot vision where the morphological transformations are especially useful for image analysis and image enhancement. The main limiting factor for the application of the morphological transformations in remotely sensed imagery, may have been the inadequate spatial resolution of satellite imagery available until recently. However, the advent of the new generation high-resolution sensors will provide the necessary resolution that will allow new algorithms, such as those based on mathematical morphology to be added to the remote sensing image processing toolkit.

3. Experimental Data

The test area used in the classification experiments was located in Athens, Greece and the data set comprised an IRS-1C panchromatic imagery from 23 April 1997 (figure 3.1). The training data required for our experiments came from a very detailed vector database recorded at the scale of 1:1000 with standard photogrammetric techniques which were applied to aerial photography acquired at a scale of 1:7000.

The reference vector dataset was first recoded in order to reduce the number of target "objects" and then rasterised at a very high spatial resolution (0.5 × 0.5 metres). In order to preserve at the maximum the structural information of the original image, the ground truth data set was referenced over the original satellite data. With this procedure the geographical reference was lost but the raster structural information that can be lost or degraded by the usual interpolation techniques was retained. The co-registration was performed by manual selection of 55 uniformly distributed ground control points with an RMS error equal to 0.27 metres. The passage from the resolution of 0.5 × 0.5 metres of the raster recoded ground truth data to the 5 × 5 metres of the satellite data set require a strong re-sampling. In order to retain the maximum of the information, we have re-sampled every different ground truth class separately with bilinear re-sampling. With this procedure, each 5 × 5 metres resolution layer relative to each ground truth class can be used as fuzzy membership function.

Fig. 3.1. Sub-image of the original IRS-1C panchromatic imagery from Athens, Greece, 24 April 1997 (©Antrix, SIE, Euromap Neustrelitz).

A set of were land use classes, representative of the majority of the land cover within the experimental area, were drawn from the reference vector dataset. The seven classes identified in this way were:

1. Small buildings (residential)
2. Large buildings (administrative, commercial, industrial)
3. Roads inside built-up areas
4. Roads outside built-up areas
5. Tips and wasteland
6. Open spaces inside built-up areas
7. Open spaces outside built-up areas

4. Image Segmentation

Watersheds is the main tool provided by mathematical morphology for contour detection and image segmentation. Watersheds were introduced in image analysis by Beucher and Lantuéjoul [2] and mathematically defined by Meyer [12] and Najman and Schmitt [14]. However, except for a few simple cases where the target object is brighter than the background or vice versa, the watershed cannot be applied directly. Generally used on gradient images, it produces a severe over-segmentation, which is difficult to overcome. The standard solution of the over-segmentation problem has been introduced by

Meyer and Beucher [13] as a strategy which involves a marker selection followed by a flooding of the relief formed by the gradient obtained from these markers. The problem is then of the marker detection and this is generally solved by a morphological filtering (geodesic closing) of the gradient image and eventually followed by a threshold of the filtered gradient.

The watershed procedure assumes that edges smaller than the structuring element and/or edges of gray level less than a given threshold are not relevant. However, in order to be able to detect the fine structures present within urban regions, small (1–2 pixels) or complex, nested regions must be retained. In this complex context, any technique based on the assumption that objects can be detected by their boundaries is very hard to apply and it often leads to results that do not have a good stability.

The approach taken here, is trying to characterise the objects by their inherent morphological characteristics, instead that through their boundaries. In this approach, an "object" is a connected region of pixels having the same morphological characteristics, that is, a "flat", "bottom" or "top" zone of the gray level function for a given set of structuring elements.

The morphological profile $\Pi(\mathbf{x})$ of an image I at point $\mathbf{x}$ is a vector $\mathbf{V}$ defined as:

$$\Pi(\mathbf{x}) = \mathbf{V}(\mathbf{x}) \begin{cases} V_i(\mathbf{x}) = \gamma_\lambda(x), & i = \lambda \ \forall \ i \in [0,\ldots,n] \\ V_i(\mathbf{x}) = \phi_\lambda(x), & i = -\lambda \ \forall \ i \in [-n,\ldots,0] \end{cases} \quad (4.1)$$

where, $\gamma_\lambda(x)$ is the structural granulometry of $I(\mathbf{x})$ resulting by successive geodesic openings, $\phi_\lambda(x)$ is the structural anti-granulometry of $I(\mathbf{x})$, resulting by successive geodesic closings, n is the maximum number of iterations and λ is the number of iterations performed using the same structuring element. Figures 4.1 and 4.2 illustrate the effect of the opening and closing morphological operators to the IRS-1C panchromatic imagery. In both cases the size of the structuring element was 9×9 pixels, which is equivalent to $\lambda = 4$.

The morphological profile residual, $\Delta(\mathbf{x})$ of $I(\mathbf{x})$ ($= V_0(\mathbf{x})$), is defined by the vector of residuals G in the $\Pi(\mathbf{x})$ as:

$$\Delta(\mathbf{x}) = \{G : G_i = |V_0(\mathbf{x}) - V_i(\mathbf{x})|, \ \forall \ i \in [-n\ldots n]\} \quad (4.2)$$

That is, $\Delta(\mathbf{x})$ is a vector with a series of *top-hat* and *inverse top-hat* (or bothat) transformations associated with every step λ of the granulometry $\gamma_\lambda(x)$ and the anti-granulometry $\phi_\lambda(x)$, respectively. Top-hat is defined as the difference between the original gray value and the value from a geodesic opening operation. The inverse top-hat is defined as the difference between the value resulting from a geodesic closing and the original gray value. Figures 4.3 and 4.4 show the resulting residual images $\Delta(\mathbf{x})$ of $I(\mathbf{x})$ from the opening and closing operations illustrated in figures 4.1 and 4.2 respectively.

Detection of Urban Features Using Morphological Based Segmentation 277

Fig. 4.1. Geodesic opening of the IRS-1C panchromatic imagery using a flat structuring element of 9×9 pixels $\Pi(x) = \gamma_4(x)$.

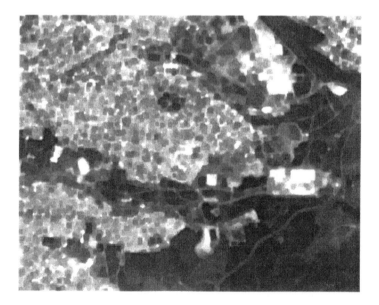

Fig. 4.2. Geodesic closing of the IRS-1C panchromatic imagery using a flat structuring element of 9×9 pixels $\Pi(x) = \phi_4(x)$.

Fig. 4.3. Residual image, resulting from figure 4.1, given by, $\Delta(\mathbf{x}) = I(\mathbf{x}) - \gamma_4(\mathbf{x})$.

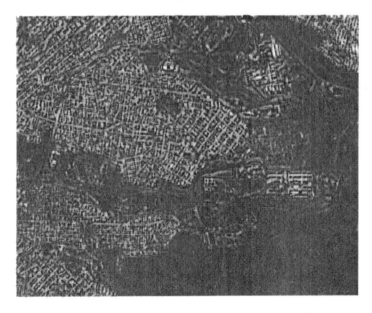

Fig. 4.4. Residual image, resulting from figure 4.2, given by, $\Delta(\mathbf{x}) = \phi(\mathbf{x}) - I(\mathbf{x})$.

Detection of Urban Features Using Morphological Based Segmentation

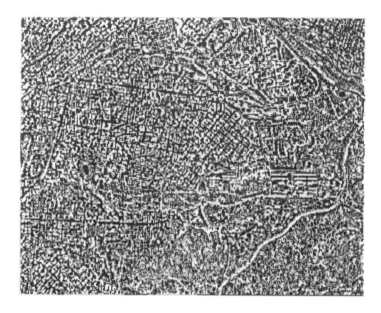

Fig. 4.5. The morphological characteristics of $\Phi(\mathbf{x})$ of $I(\mathbf{x})$. The "bottom" and "top" regions are darker and brighter respectively.

The morphological characteristics $\Phi(\mathbf{x})$ of $I(\mathbf{x})$ are then defined as:

$$\Phi(\mathbf{x}) = \begin{cases} i : \Delta_i(\mathbf{x}) = \vee(\Delta(\mathbf{x})) \text{ with } \Delta_i(\mathbf{x}) \neq \Delta_j(\mathbf{x}) \ \forall i,j \in [-n,\ldots,n] \\ 0 \text{ if } \exists \ i,j \mid \Delta_i(\mathbf{x}) = \Delta_j(\mathbf{x}) \end{cases}$$

(4.3)

An object in that sense is a set of connected pixels (region) with the same value of $\Phi(\mathbf{x})$, where, $\Phi(\mathbf{x}) \in [1,\ldots,n]$ in the case of prevalently "top regions", $\Phi(\mathbf{x}) \in [-n,\ldots,-1]$ in the case of prevalently "bottom" regions and $\Phi(\mathbf{x}) = 0$ in the case of prevalently flat or morphologically indifferent regions for all sizes of the structuring element. This is illustrated in figure 4.5 where the "top" and "bottom" regions appear bright and dark respectively.

The image of figure 4.6 shows a portion of the image test data set segmented by the morphological characteristics function $\Phi(\mathbf{x})$ and then cleaned by a local majority filter, in order to reduce the total number of regions by reducing the number of the low contrast one-pixel-wide regions.

5. Region Based Classification of the Morphological Channels

The classification of the regions resulting from the morphological operations was carried out using a multi-layer perceptron neural network trained with

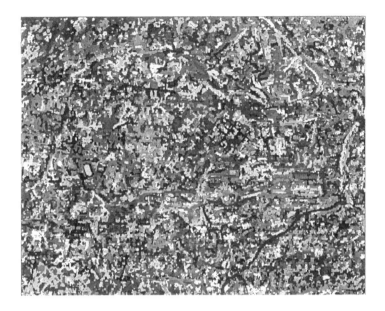

Fig. 4.6. The segmented IRS-1C image, due to the morphological characteristics function $\Phi(\mathbf{x})$ – figure 4.5.

the back-propagation algorithm. A neural network classifier was selected because the combined features resulting from the various morphological operations are not likely to obey a good statistical model. Furthermore neural networks have proven to be very robust for mixed or multi-source data sets where it is not easy to describe the data by a fixed statistical model [1, 7].

The multi-layer perceptron used in the experiments consisted of 3 layers. The number of nodes in the input layer varied according to the number of features used in the experiments. The number of nodes in the hidden layer were determined experimentally and in each case we selected the architecture that resulted in the best class accuracies. The output layer consisted of seven nodes, i.e. the number of land cover/land use classes in the training data. In total 2621 regions were used to train the neural network and 2576 regions to verify the performance of the classifier.

Five different classification experiments were performed, using different data sets as illustrated in Table 5.1. This table gives the map producer's accuracies for each class, for each data set employed. The accuracies have been computed on the verification data sets. In order to assess the value of the features provided by the morphological channels a neural network was also trained to classify the 7 land use classes using only the means and standard deviations computed from the original radiometric values of the IRS-1C panchromatic imagery together with the area of each region and two shape descriptors resulting from the two major axes of each region (DataSet

Detection of Urban Features Using Morphological Based Segmentation 281

Table 5.1. Classification accuracy comparisons for different data sets. DS1 = original panchromatic image values + area + 2 shape descriptors Data Set, DS2 = DS1 + means and standard deviations of regions resulting from all the opening and closing operations, DS3 = DS2 but without the area feature, DS4 = DS1 + means and standard deviations of regions resulting after the $1^{st}, 2^{nd}$ and 3rd opening and closing operations and DS5 = DS2 but without the area feature.

Class Name	Per-Class Map Producer's (%) Accuracies				
	DS1	DS2	DS3	DS4	DS5
Small buildings	80.22	81.13	77.50	82.03	79.67
Large buildings	33.33	46.67	86.67	66.67	66.67
Roads inside built-up areas	59.26	70.99	70.99	69.34	73.66
Roads outside built-up areas	31.25	71.88	82.81	73.44	78.12
Tips and wasteland	28.57	100.0	100.0	90.48	100.0
Open spaces inside built-up areas	2.50	67.50	67.50	47.50	50.00
Open spaces outside built-up areas	99.50	99.50	99.29	99.36	99.36
Overall Accuracy	83.62	88.70	88.32	88.28	88.82

1 - DS1). The overall classification accuracy achieved with DS1 was 83.62%, with only two classes "Small Buildings" and "Open spaces outside build-up areas" achieving accuracies above 80%. The rest of the classes are classified with accuracies less than 60%, while the class "Open spaces inside build-up areas" was the least accurately classified at only 2.5%.

The addition of the features resulting from the morphological opening and closing operations improved the overall accuracy by 4.66% to 88.24% when DS4 (like DS1 but with the addition of the means and standard deviations of the regions, from only 3 opening and closing operations - resulting in 17 input features to the network) is used. With DS2, which includes the features from all the morphological opening and closing operations, i.e. 41 input features to the neural network, the classification accuracy was further improved by 5.1% to 88.70%. Therefore the addition of the morphological features has yielded an overall improvement in the discrimination of the seven land use classes. Although the overall gain may not seem large the gains at the individual class level are more significant as revealed in table 5.1. For example through the use of DS2 there is an improvement of 65% for the worst class ("Open spaces inside build-up areas"), and gains of 71.4% for "Tips and wasteland" and 40.6% for "Roads outside build-up areas". Similar improvements are obtained also through the use of DS4. The classification accuracies of DS2 and DS4 are comparable, however the addition or exclusion of some morphological features affects the accuracies of "Large buildings" and "Open spaces inside build-up areas".

It is also interesting to note the results obtained, when the measure of the area of each region is excluded from the input features. Datasets DS3 and DS5 are identical to DS2 and DS4 respectively, but without the area measure. From table 5.1 we can see that this may have a significant positive or negative contribution to the accuracy of buildings and roads. In particular in both cases there is a significant improvement in the classification of "Roads outside

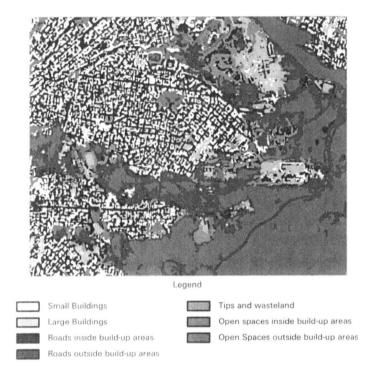

Fig. 5.1. Classification of the IRS-1C imagery from Athens into seven man-made classes.

build-up areas", ranging from 4.7% for DS5 to 10.9% for DS3. A substantial improvement, of 40%, was also obtained for the "Large buildings class" with DS3. In both cases however, the classification accuracy of "Small buildings" has been reduced by a small amount. Figure 5.1 shows the classification of the Athens IRS-1C imagery, based on a neural network trained with the DS3 dataset.

6. Discussion

In the experiments presented in this chapter it is demonstrated that significant gains can be made in applying transformations from mathematical morphology to very-high resolution imagery for the segmentation and classification of man-made areas. In particular morphological features were extracted from IRS-1C imagery through successive morphological geodesic opening and closing operations. The morphological characteristics were then used to segment the image. A multi-layer perceptron neural network was subsequently trained, using information derived from the resulting segments, to classify the imagery into seven man-made classes. It has to be noted that only the panchromatic channel of the IRS-1C imagery was used in all the above operations.

Overall it was shown that it is possible to extract seven man made classes with a good accuracy. However these classes are unavoidably "fuzzy" classes and our experiments reveal that there is a lot of confusion between the "Small buildings" and the "Roads inside build-up areas" classes. Furthermore it seems that inclusion or omission of some of the features used may have significant effects in the classification accuracy of some of the classes. Currently we are investigating both fuzzy morphological procedures and feature extraction techniques that may further aid the discrimination of man-made areas.

References

1. J. A. Benediktsson, J. R. Sveinsson, and K. Arnason, "Classification and feature extraction of AVIRIS data", *IEEE Transactions on Geoscience and Remote Sensing*, vol. 33, no. 5, pp. 1194–1205, 1995.
2. S. Beucher and C. Lantuéjoul, "Use of watersheds in contour detection", in *International Workshop on Image Processing, CCETT/IRISA, Rennes, France*, September 1979.
3. J. Crespo, J. Serra, and R. Shafer, "Theoretical aspects of morphological filters by reconstruction", *Signal Processing*, vol. 47, pp. 201–225, 1995.
4. G. Flouzat, O. Amram and S. Cherchali, "Spatial and spectral segmentation Of satellite remote sensing imagery using processing graphs by mathematical morphology", *Proceedings, Geoscience and Remote Sensing Symposium, IGARSS'98*, 6-10 July 1998, Seattle, WA, vol. IV, pp. 1769–1771, 1998.
5. H. J. A. M. Heijmans, "Mathematical morphology as a tool for shape description", in *Shape in Picture: Mathematical Description of Shape in Grey-Level Images*, 7–11 September 1992, Driebergen, The Netherlands, pp. 147–176, 1992.
6. B. Jähne, *Digital Image processing*, Berlin:Springer-Verlag, 1991.
7. I. Kanellopoulos, G. G. Wilkinson, and A. Chiuderi, "Land cover mapping using combined Landsat TM imagery and textural features from ERS-1 Synthetic Aperture Radar Imagery", in *Image and Signal Processing for Remote Sensing* (Jacky Desachy, ed.), Proceedings SPIE 2315, pp. 332–341, September 1994.
8. C. Lantuéjoul and S. Beucher, "On the use of the geodesic metric in image analysis", *Journal of Microscopy*, vol. 121, pt. 1, pp. 39–49, 1981.
9. P. Le Queré, P. Maupin, R. Desjardins, M.-C. Mouchot, B. St-Onge and B. Solaiman, "Change detection from remotely sensed multi-temporal images using morphological operators", *Proceedings, Geoscience and Remote Sensing Symposium, IGARSS'97*, 3–8 August 1997, Singapore, vol. I, pp. 252–254, 1997.
10. G. Matheron, *Random sets and integral geometry*. New York: Wiley, 1975.
11. P. Maupin, P. Le Queré, R. Desjardins, M.-C. Mouchot, B. St-Onge and B. Solaiman, "Contribution of mathematical morphology and fuzzy logic to the detection of spatial change in an urbanised area: towards a greater integration of image and Geographical Information Systems", *Proceedings, Geoscience and Remote Sensing Symposium, IGARSS'97*, 3–8 August 1997, Singapore, vol. I, pp. 207–209, 1997.

12. F. Meyer, "Integrals, gradients and watershed lines", in *Mathematical Morphology and its application to Signal Processing*, Symposium held in Barcelona, pp. 70–75, May 1993.
13. F. Meyer and S. Beucher, "Morphological segmentation", *Journal of Visual Communication and Image Representation*, vol. 1, no. 1, pp. 21–46, 1990.
14. L. Najman and M. Schmitt, "Definition and some properties of the watershed of a continuous function", in *Mathematical Morphology and its application to Signal Processing*, Symposium held in Barcelona, pp. 75–81, May 1993.
15. P. L. Palmer and M. Petrou, "Locating boundaries of textured regions", *IEEE Transactions on Geoscience and Remote Sensing*, vol. 35, no. 5, pp. 1367–1371, 1997.
16. P. Salembier and J. Serra, "Multiscale image segmentation", in *Visual Communication and Image Processing*, vol. 1818, pp. 620–631, 1992.
17. J. Serra, *Image analysis and mathematical morphology*. London: Academic Press, 1982.
18. J. Serra, "Introduction to mathematical morphology", *Computer Vision Graphics and Image Processing*, vol. 35, pp. 283–305, 1986.
19. J. Serra, ed., *Image analysis and mathematical morphology*. Vol. 2, Theoretical Advances, New York: Academic Press, 1988.
20. J. Serra and P. Salembier, "Connected operators and pyramids", in *Non-Linear Algebra and Morphological Image Processing*, vol. 2030, pp. 65–76, 1993.
21. L. Vincent, "Graphs and mathematical morphology", *Signal Processing*, vol. 47, pp. 365–388, 1989.
22. D. Wang, D. C. He and D. Moran, "Classification of remotely sensed images using mathematical morphology", *Proceedings, Geoscience and Remote Sensing Symposium, IGARSS'94*, 8–12 August 1994, Pasadena, CA, vol. III, pp. 1615–1617, 1994.

Non-Linear Line Detection Filters

Isabelle Gracia[1] and Maria Petrou[2]

[1] Departamento de Informática, Universidad Jaume I,
12071 Castellón, Spain.
e-mail: gracia@uji.es
[2] School of Electronic Engineering, Information Technology and Mathematics,
University of Surrey, Guildford, GU2 5XH, UK.
e-mail: m.petrou@ee.surrey.ac.uk

Summary. We present here a set of non-linear filters appropriate for the enhancement of linear features in remote sensing images. The particular features we are interested in enhancing and identifying are buried underground structures that might be of interest, for example, to Archaeology, or large scale surveys. Our results are demonstrated with the help of grey-scale 1m resolution aerial images.

1. Introduction

Linear structures in an image have two major differences from the edges that are usually the subject of edge-enhancing filter research: They are symmetric features (i.e., they have cross-section profiles that are symmetric about their centres in the idealised forms) and they have intrinsic length scales (i.e., they have finite widths that determine the size of the filter that would be best for them) [2]. Some linear filters have been proposed in the past for the detection of linear structures in preference to edge type features. Those filters have been proven very successful in identifying roads, hedges, etc. in remote sensing images [2, 3]. However, they can only cope with linear structures that have a certain contrast and regularity with respect to the surrounding regions. Figure 1.1b shows an image that contains a buried linear structure. A black line that represents the central position of this buried linear structure is shown in figure 1.1(c). This structure manifests itself by influencing the over-ground appearance of the vegetation and the soil, but its signal is so weak, that no linear filter is expected to be able to cope with it. In fact, the image in figure 1.1(b) has been processed by histogram equalisation for presentation purposes. The buried linear structure is hardly visible in the raw image shown in figure 1.1(a). All the experimental results presented in sections 4. and 5. were performed using the raw images. The images are all of size 256×256, and correspond to various land covers, including ploughed fields and grasslands.

In this chapter we propose a series of non-linear filters, some of which are modified versions of those proposed by Pitas and Venetsanopoulos [4] for edge detection. We then apply them to enhance the relevant lines in an image. The final line detection is performed with the help of Hough Transform [1].

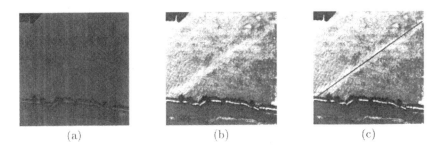

Fig. 1.1. (a) Image that contains a buried linear structure, (b) Image after histogram equalisation, (c) The linear structure identified manually.

2. The Triptych Window

Linear features in an image manifest themselves as maxima in the second derivative of the image function. Further, due to their "symmetric" cross-section, the ideal matched filter for them would be one with symmetric profile. That is why strong linear features can be enhanced with the help of "Sombrero"-type filters. Statistical non-linear filters used for the same job must follow similar guidelines. So, we introduce the concept of a triptych window.

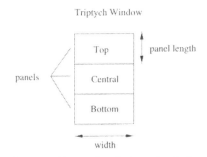

Fig. 2.1. Example of triptych window.

A triptych window is a window composed of three panels, as shown in figure 2.1. Then we will compute the following equation to express the filter response corresponding to the specified triptych window:

$$F(P) = F_{Central}(P) - \frac{F_{Top}(P) + F_{Bottom}(P)}{2}$$

where $F(P)$ is the global response of the filter F when the central point of the triptych window is at position P, and $F_{Central}(P)$, $F_{Top}(P)$, and $F_{Bottom}(P)$ are the responses obtained for each panel of this triptych window.

As we are interested in obtaining just one filter response (peak) in the described position, we must set a length for the central panel as long, at least, as the linear structure width. Otherwise, we would obtain two (wrong) responses, first when the triptych window is getting into the linear structure and second when is leaving it.

3. Filters Investigated

Let N be the number of pixels in a panel of a triptych window, and x_i the discrete brightness level of pixel i, where $i = 1, 2, \ldots, N$. In order to detect the desired linear structure, we applied the following statistical filters:

- Mean, defined as $\bar{x} = \left(\sum_{i=1}^{N} x_i\right)/N$
- Variance, defined as $\sigma^2 = \left(\sum_{i=1}^{N} (x_i - \bar{x})^2\right)/N$
- Trimmed Mean. It is obtained by recalculating the mean value of the pixels inside an interval centred at $\bar{x}$ and having width 2σ, that is, using only the pixels that have values in the range $[\bar{x} - \sigma, \bar{x} + \sigma]$. This statistic is more robust than the mean.
- Median. It is obtained by ranking the values x_i, and taking the central value when the number N of elements in the panel is odd, and the mean of the two central values when N is even.
- Mode, defined as the most frequent value x_i in the panel.
- Entropy, defined as $H = -\sum_{i=1}^{N} P(x_i) \cdot \log P(x_i)$, where $P(x_i)$ is the probability of having value x_i in the panel.
- Energy, defined as $E = \sum_{i=1}^{N} P(x_i)^2$, where $P(x_i)$ is defined as before.

In our experiments, we computed the entropy and the energy using binned data. This means that we have modified the general equations given above in the following way:

$$H_b = -\sum_{i=1}^{B} P(bin_i) \cdot \log P(bin_i)$$

$$E_b = \sum_{i=1}^{B} P(bin_i)^2$$

where B is a free parameter that specifies the number of bins to use, and $P(bin_i)$ is the probability of having a discrete brightness value belonging to bin i of the corresponding panel. To obtain the bins for each panel of the triptych window, we computed the minimum and maximum discrete brightness values for the whole window, and divided this interval in the number of desired bins. The reason we did that is in order to have more reliable statistics as the number of pixels inside a panel is rather small, typically a few dozen.

Table 3.1. Nonlinear means.

Arithmetic	Geometric	Harmonic	L_p $(p \in Q - \{1, 0, -1\})$	Contraharmonic Ch_p
$\dfrac{\sum_{i=1}^{N} a(i) x_i}{\sum_{i=1}^{N} a(i)}$	$\exp\left(\dfrac{\sum_{i=1}^{N} a(i) \log x_i}{\sum_{i=1}^{N} a(i)}\right)$	$\dfrac{\sum_{i=1}^{N} a(i)}{\sum_{i=1}^{N} \frac{a(i)}{x_i}}$	$\sqrt[p]{\dfrac{\sum_{i=1}^{N} a(i) x_i^p}{\sum_{i=1}^{N} a(i)}}$	$\dfrac{\sum_{i=1}^{N} x_i^{p+1}}{\sum_{i=1}^{N} x_i^{p}}$

Table 3.2. Order statistics.

Range	Dispersion
$w_{(1)} = x_{(N)} - x_{(1)}$	$W_{N/2} = \sum_{i=1}^{N/2} w_{(i)} = \sum_{i=1}^{N/2} x_{(N+1-i)} - x_{(i)}$

We also applied all filters described in [4]. These filters are based on nonlinear means and on order statistics. The definition of nonlinear means is shown in table 3.1, where p is a free parameter for computing the L_p mean and the contraharmonic mean. Quantities $a(i)$, $i = 1, \ldots, N$, are user defined weights. The definition of the order statistics filters is given in table 3.2. In this case, $x_{(i)}$ is the order (i) value of the pixels of the panel. Thus, $x_{(i)}$ represents the discrete brightness level of a pixel once all the pixels in the panel have been arranged in increasing order, that is, $x_{(1)} \leq x_{(2)} \leq \ldots \leq x_{(N)}$.

4. Experiments to Choose the Right Filter

We carried out several experiments in order to identify the most appropriate filters for the buried linear structure detection problem. Figure 4.1 shows the raw images used in the experiments. These images are 256 × 256 pixels in size, and correspond to grey-scale 1m resolution aerial images. Figure 4.2 shows the ground truth of these images (after equalising the histogram for displaying purposes), that is, the true position of the buried linear structure we want to detect. To compare the performance of the various filters, several experiments were performed by scanning these images along various lines orthogonal to the true direction of the lines to be detected. Figure 4.3 shows some example filter responses obtained from test image 1. These responses have been smoothed with a Gaussian mask with $\sigma = 2.3$, so that only the significant peaks remain. In each case the true location of the line to be detected is at position 0. Note that since here we are only interested in identifying the best filters, we scan

Non-Linear Line Detection Filters   289

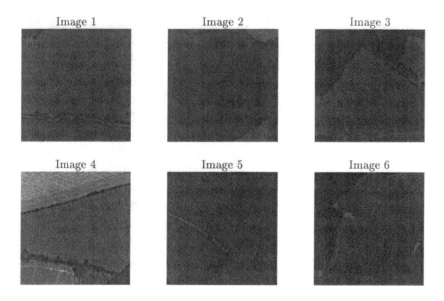

Fig. 4.1. Images used in the experiments.

the image along each chosen direction for a short length only on either side of the line to be detected.

To quantify the obtained results in a systematic way in order to compare the performance of the studied statistical filters, we define the following parameters:

- Q_1. It measures the discrimination between the two highest peaks in the filter response:

$$Q_1 \equiv \frac{|peak_1| - |peak_2|}{|peak_1|}$$

where $peak_1$ and $peak_2$ denote the highest and the second highest peak in the filter output respectively.
- Q_2. It measures the distance between the position of the highest peak in the output and the true position of the linear structure.

A good filter response should maximise Q_1, that is, the highest peak should be as clearly identifiable as possible, and minimise Q_2, that is, the peak should be as close as possible to the true location of the linear structure.

After carrying out several experiments using several points in all images shown in figure 4.1 and studying the values Q_1 and Q_2 obtained, we drew the following conclusions:

1. When there are interfering lines, that is, there are other edges close to the desired linear structure, the best results were obtained with the mode,

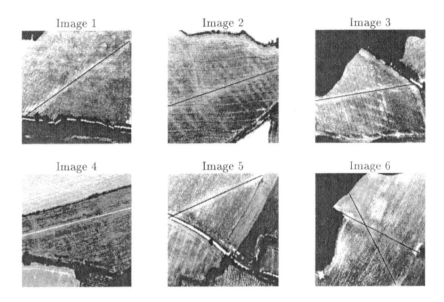

Fig. 4.2. Images used in the experiments. Black lines that represent the buried linear structures have been drawn, except for image 4, where a white line has been used.

the median, and the trimmed contraharmonic filters. This is not really surprising as these filters are robust filters, especially designed to deal with interference.

2. All robust filters (i.e., trimmed contraharmonic, trimmed mean, median and mode) gave also good results in those cases that do not suffer from interference.
3. For the images with no interference, the harmonic, geometric, arithmetic, L_p and contraharmonic means gave also good results.
4. The variance, range, the energy, and the entropy gave systematically the worse results, while the dispersion seems to behave in an erratic way.

5. Detecting Underground Structures

In order to locate faint linear structures without using a priori knowledge about their positions, we carried out the following experiment, using one of the best filters identified in the previous section, namely the median. First, we computed the response of the statistical filter when the triptych window was moved horizontally and then vertically, along each row and column of the image separately. To enhance the most significant peaks, we convolved each output signal with a Gaussian mask. Looking at the filter responses of figure 4.3, where the panel length in the triptych window was fixed to 18 pixels, we

Non-Linear Line Detection Filters 291

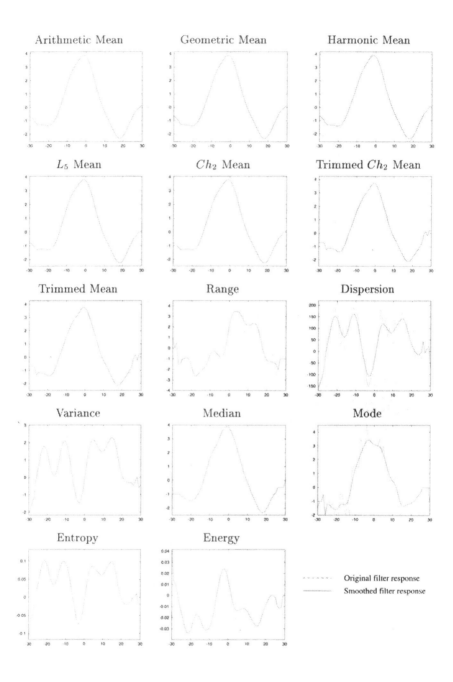

Fig. 4.3. Responses of the various statistical filters.

notice that a buried line manifests itself as a maximum of the filter response surrounded from both sides by two minima at distance approximately equal to the length of the panel of the window. We also expect that, in an ideal case, the magnitude of the two minima should be half of the maximum value. So, for every extremum in the smoothed output signal at place p, we checked if there were two extrema of the opposite sign in symmetric positions, at distances $d \pm \delta d$ from p, where d is the panel length of the triptych window used, and δd is some tolerance. Furthermore, if the output signal for the candidate extremum is V_p and for its neighbouring extrema is V_{n_1} and V_{n_2}, we also check if $V_{n_1}, V_{n_2} \in [-V_p/2 \pm \gamma V_p]$, where γV_p is some tolerance. If the candidate point satisfies these two requirements it is accepted as an edge point, otherwise it is rejected.

Figure 5.1(b) shows the response of the filter along the column of the image 3 shown in figure 5.1(a). The vertical line in figure 5.1(b) marks the true position of the buried linear structure, and it is also the only selected peak in the graph. The deep minimum followed by the strong maximum on the left is the signature of the step edge with the black region on the top of the image. The true location of the step edge is at the zero crossing between these two extrema. This shows that picking actually the extrema of the filter response according to their magnitude only and not using the criterion of symmetric negative responses on either side, would lead to wrong results.

 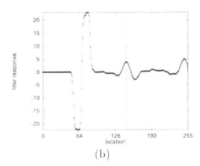

(a) (b)

Fig. 5.1. (a) The white line marks the column used in the experiment, (b) Response of the filter along the selected column. The vertical line marks the true position of the desired linear structure.

Figure 5.2 shows the results obtained after applying the triptych window along horizontal and vertical lines. The first column shows the original images. The second column presents the selected points that, after the horizontal and vertical movement of the window, satisfied all symmetry and magnitude requirements. The images on the third column display the main linear structure detected applying the Hough transform to the selected points. These grey images are after histogram equalisation.

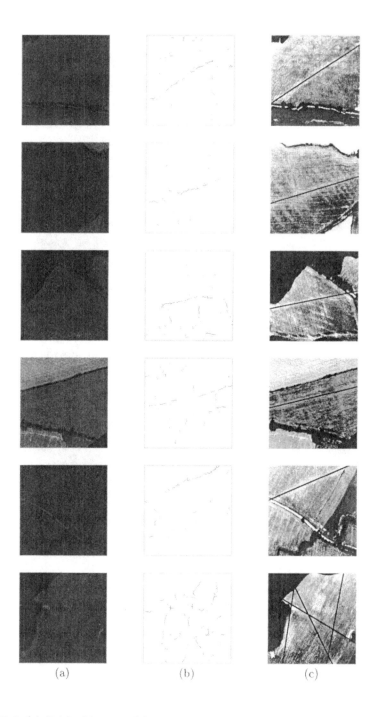

Fig. 5.2. (a) Original images, (b) Edges obtained after applying the triptych window to every column and row in the raw image, (c) Lines identified by the Hough transform.

All these results were obtained with the same set of parameters, namely panel width of 21 pixels, panel length 18 pixels, $\delta = 0.3$ and $\gamma = 0.4$. Also the parameters for the Hough transform were fixed and each time only the strongest peak in the parameter space was chosen. For the image with the two crossing lines (test image 6), we have to pick the 4 strongest peaks in Hough space in order to identify both crossing lines, as one of the two is enhanced in a very fragmented way. The Hough transform used is the one described in [1].

6. Conclusions

We presented here non-linear filters appropriate for enhancing very noisy linear structures, like those observed in remote sensing images with Archaeological interest. The best filters are those based on robust statistics because such images contain significant interference from over-ground linear structures. In particular, filters based on the trimmed contraharmonic, the mode and the median performed best.

Although these filters are most appropriate for detecting lines when some prior knowledge on the true location of the line is available, we have demonstrated here that they can be used even with no prior knowledge available, as directional filters that are applied along two orthogonal directions of the image, the results of which can be combined to enhance the structures they are tuned to. Tuning of these filters entails choosing the right size of each panel of the triptych window. This is an important parameter as it allows us to enhance preferentially structures of selected width.

Acknowledgement. This work was partly supported by the British National Space Agency, project ADP3, and a Fundacio Caixa-Castello grant.

References

1. H. Kalviainen, P. Hirvonen, and E. Oja, "Houghtool—a software package for hough transform calculation", in *Proceedings of the 9th Scandinavian Conference on Image Analysis, Uppsala, Sweden*, vol. 2, pp. 841–848, June 1995.
2. M. Petrou, "Optimal convolution filters and an algorithm for the detection of wide linear features", *IEE Proceedings I. Communications Speech Vis.*, vol. 140, no. 5, pp. 331–339, 1993.
3. M. Petrou, P. Papachristou, and J. Kittler, "Probabilistic-relaxation-based postprocessing for the detection of lines", *Optical Engineering*, vol. 36, no. 10, pp. 2835–2844, 1997.
4. I. Pitas and A. N. Venetsanopoulos, "Edge detectors based on nonlinear filters", *IEEE Transactions on Pattern Analysis and Machine Intelligence*, vol. PAMI-8, no. 4, pp. 538–550, 1986.

Fuzzy Clustering and Pyramidal Hough Transform for Urban Features Detection in High Resolution SAR Images

Eugenio Costamagna, Paolo Gamba, Giacomo Sacchi, and Pietro Savazzi

Dipartimento di Elettronica, Università di Pavia,
Via Ferrata, 1, I-27100 Pavia, Italy.

Summary. In this work a fuzzy version of the Connectivity Weighted Hough Transform (CWHT) is used to detect streets and roads in high resolution (synthetic Aperture Radar (SAR) images of urban environments. The basic idea is to define a pyramidal procedure suitable for the characterization of several types of streets having different width. In this sense a modified Hough Transform, based on results produced by fuzzy clustering techniques, is introduced to group pixels classified as possible roads in consistent straight lines.

1. Introduction

One of the most interesting research topics in the analysis of remotely sensed imagery is connected with the fusion of Synthetic Aperture Radar (SAR), hyperspectral, multispectral and aerial photo images to better detect and characterize different urban environments [6]. However, there still do not exist reliable analysis methods, especially when considering the difficulties arising from the use of radar data into urban areas. The dependency of the backscattered field from the geometric as well as dielectric characteristics of the man-made structures makes the retrieved radar images very noisy. Therefore, even if there is a large number of publications that consider the relevance and usefulness of multiband and/or multipolarization radar data (see for instance [5, 7]), many works now explore possible automatic classification methods for different structures [4, 8, 13, 16]. Among them, we refer in particular to those aimed to street detection [3, 9, 10, 11] since this is clearly a problem where remote sensing and machine vision areas overlap. In this sense we observe that many of the methodologies used by the image processing community (Bayesian restoration, Markov fields, Hough transform, to name only a few) are commonly applied to satellite and aerial images, often exploiting results used first in standard image interpretation.

In this chapter we introduce these methods in the analysis of high resolution radar data, that are now increasingly available and promise to provide a rich information source in remote sensing applications(3D building extraction, for instance, see [12]). In particular, in this chapter we shall show a way to classify SAR data with machine vision approaches suitable (in an iterative fashion) to extract features with great precision. To achieve this goal, a useful approach is first to study the classification problem as a clustering one, introducing for instance standard fuzzy clustering techniques [1]. It is well-known that Fuzzy C-Means (FCM) or Possibility C-Means (PCM) algorithms are useful for clustering data, and especially the latter is less noise-sensitive (therefore, more reliable in radar image analysis). With this approach we avoid to classify immediately all the urban environments, and we concentrate first in a very rough characterization of the road network, the green areas and the built zones.

Then, the data classified by the previous step could be considered as an input of more refined algorithms for model extraction. This is the case, again, when aggregating pixels recognized as streets or roads into consistent lines. Therefore, assuming the fuzzy membership function computed in the FCM or PCM clustering step, we implemented a fuzzy clustering procedure aimed to extract straight lines from the original image by weighting the role of each pixel by its membership to the street class.To do this we modified the Connectivity Weighted Hough Transform (CWHT) presented in [14] by considering data that is not black and white, but it has some sort of grey level based on the corresponding fuzzy membership value.

The usefulness of the proposed approach is validated by using high resolution satellite data of the centre of Milan, Italy, and furthermore analyzing some AIRSAR images of Santa Monica, California. All the classification chain was implemented by means of a simple PC program, with very friendly interface, called EAGLE.

The chapter is organized as follows: Section 2 presents the complete procedure, introducing the clustering algorithm, used in this analysis, while section 3 is devoted to the novel fuzzy pyramidal CWHT approach to detect straight lines with different thickness (see figure 1.1). Finally, section 4 shows the results obtained on a C-band SAR image of Los Angeles and an IRS-1 panchromatic image of Milan (Northern Italy). Section 5 concludes the chapter discussing the proposed method as well as propose new research directions.

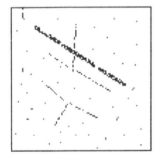 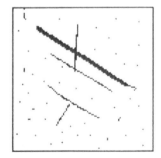

Fig. 1.1. An example of lines with different thicknesses before and after the classification step.

2. The Detection Algorithm

Keeping in mind the ideas outlined in the introduction, we applied a conceptually pyramidal approach, avoiding to classify immediately all the different classes that an urban environment contains, and concentrating on a very rough characterization of the road network, the green areas and the built zones. A more refined analysis is successively applied only to these preliminary results, in order to extract as much as possible information from the radar image, for instance by clustering road pixels into consistent straight or curvilinear lines, and further subdividing the built areas into different land-use zones. The present work focuses especially on the capability of a Pyramidal Fuzzy CWHT algorithm for the extraction of straight lines as street pixels from the raw data.

2.1 The FCM and PCM Clustering Algorithms

The first processing step is devoted to the analysis of the original data by means of the FCM and PCM techniques. It is well known that both of these approaches [1] can be reduced to a generic clustering of the data with the difference that the FCM does not consider explicitly a class for noise data (i.e. image grey values that do not correspond to any of the searched classes) while PCM considers a noise class for each of the clusters in which the original image is subdivided.

This becomes clear by considering the functional minimized by an iterative procedure in the two methods:

$$J_{FCM} = \sum_{i=1}^{c} \sum_{j} u_{ij}^m d^2(x_j, \epsilon_i)$$

$$J_{PCM} = \sum_{i=1}^{c} \sum_{j} u_{ij}^m d^2(x_j, \epsilon_i) + \sum_{i=1}^{c} \sum_{j} (1 - u_{ij}^m)$$

where x_j is the j-th element of the image, c is the number of clusters, $d(\cdot)$ is some sort of distance measure (for instance, the Mahalanobis distance), ϵ_i is i-th cluster centre and $\{\eta_i\}$ are weights that can be chosen to determine how important is the noise class for each data cluster. The membership function u_{ij} is computed by using $m \in [1, 2]$ and:

$$u_{ij}^{FCM} = \frac{1}{\sum_{k=1}^{c} \left(\frac{d^2(x_j, \epsilon_i)}{d^2(x_j, \epsilon_k)}\right)^{\frac{1}{m-1}}}$$

$$u_{ij}^{PCM} = \frac{1}{\sum_{k=1}^{c} \left(\frac{d^2(x_j, \epsilon_i)}{\eta_i}\right)^{\frac{1}{m-1}}}$$

Both methods can be used for an unsupervised classification of an image and it is well known that PCM is, by definition, more robust to noise than

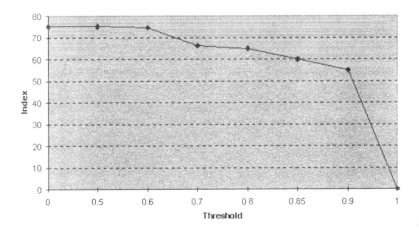

Fig. 2.1. A typical behaviour of the confidence index i with respect to the classification threshold applied to the fuzzy clustered data.

FCM. For this reason in our classification chain a preclustering by means of FCM algorithm provides suitable η values for the effective PCM unsupervised classification of the SAR image. However, to do this it is necessary to extract a thematic map from the data, i.e. to define a classification threshold to be applied to the fuzzy clustered results. This threshold was chosen by looking at the confidence index (one for each class),

$$i = \frac{G}{G + B_l + B_c + N_c}$$

where G is the number of pixels correctly assigned to the class, $B - l$ is the number of pixels erroneously assigned to the same class, $B - c$ is the number of pixels assigned to other classes and finally $N - c$ is the number of pixels of the considered class that were not classified. A typical situation is depicted in figure 2.1.

3. The Fuzzy Pyramidal CWHT for Street Detection

To refine the analysis, the data classified by the previous step may be considered as an input of more target oriented algorithms for model extraction. This is the case, for instance, when aggregating pixels recognized as streets or roads into consistent lines.

In this work, we assume not to lose the fuzzy confidence levels (i.e. the clustering resulted membership function) computed in the PCM (or FCM) clustering step. Instead, these values must be fully considered in all subsequent processing, so that no information is lost. Therefore, we implement a fuzzy pyramidal procedure aimed to extract straight lines from the original image by weighting the role of each pixel by its membership to the street class. The Connectivity Weighted Hough Transform (CWHT), presented in [14] and similar to the approach formulated in [15], was modified by starting from data that is not black-and-white (like all the examples in [14]), but has some sort of grey level based on the corresponding fuzzy membership value.

The CWHT algorithm relies on the use of the connectivity among pixels to be grouped into the same segment, as in [15]. By exploiting this property, a dramatic acceleration in the Hough Transform (HT) computing time is obtained, since the HT search space is somehow limited by the fact that the pixels grouped into a segment must be connected; this idea leads also to a straightforward determination of the start and the end pixels of each segment, i.e. a processing step further than the standard HT. The CWHT widens this definition by taking into account also the line thickness, and considering connectivity not only in the segment's direction but also along its width. The practical implementation of this technique is based on a likelihood principle of connectivity and thickness.

Consider one of the many ways that a straight line can be represented, e.g. the Duda and Hart's polar equation [2]

$$x \cos \theta + y \sin \theta = \rho$$

where θ and ρ are the two parameters of the line, having finite range. For each fixed point (x_i, y_i), the above polar equation become a curve in the (ρ, θ) plane. When all the n curves are plotted in the (ρ, θ) plane, the intersection between most members of the curves may represent a line that has the highest number of collinear points. In the traditional approach of Hough Transform, in a noisy image it is difficult to recognize a short thick line segment that can instead be easily recognized by human eyes; the measure of connectivity can alleviate this problem. Furthermore, in digital images, due to quantization, exact co-intersection at one point by many (ρ, θ) curves may not be possible. Thus the (ρ, θ) plane is usually divided into many rectangular cells

$$R_{ij} = [\rho_i, \rho_{i+1}) \times [\theta_i, \theta_{i+1})$$

with $i = 1, \ldots, C_\rho, j = 1, \ldots, C_\theta$, where C_ρ and C_θ are pre-determined class sizes, and $\{\rho_i\}$ and $\{\theta_j\}$ are monotonic sequences so that all the cells cover the whole range of ρ and θ.

Let the prototype line thickness be d and n_{ij} be the number of pixels on the line segment connecting the image points P_i and P_j, i.e.

$$n_{ij} = \text{all the pixels that have distance} \leq d/2 \text{ to the line segments connecting } P_i \text{ and } P_j$$

Let x_{ij} be the number of street pixels in these n_{ij} pixels. Then, we can define the weight that can be used to decide if the points P_i and P_j are connected by a line of thickness d for the cell R_{ij}, as.

$$\omega_{ij} = \begin{cases} x_{ij} \log\left(\frac{x_{ij}}{n_{ij}p_0}\right) + \left(n_{ij} - x_{ij} \log\left(\frac{n_{ij}-x_{ij}}{n_{ij}(1-p_0)}\right)\right) & \text{if } x_{ij}/n_{ij} > p_0 \\ 0 & \text{otherwise} \end{cases}$$

with the convention $0 \ln(0) = 0$. This formula derives from the following likelihood ratio,

$$L = \frac{p^{x_{ij}}(1-p)^{n_{ij}-x_{ij}}}{p_0^{x_{ij}}(1-p_0)^{n_{ij}-x_{ij}}}$$

where p is the probability of a pixel being street given that it is on a line. The maximum likelihood algorithm estimates $\hat{p} = x_{ij}/n_{ij}$ and decides for the presence or absence of a straight line in the cell R_{ij}. In both fuzzy and pyramidal approach we have considered at each pyramidal level straight lines having decreasing thickness d and we have defined as x_{ij} the sum of membership functions, provided by fuzzy clustering, of the n_{ij} pixels, classified as road pixels, to the street class. From the coarsest to the most precise level we adopted decreasing values for the line thickness following the formula

$$d(L) = \frac{w}{(N_L - 1)}(L - 1) + 1$$

where N_L is the implemented pyramidal level number, L represents the current layer under examination (its range is $1, \ldots, N_L$), while w is the maximum line thickness plus 1 corresponding to the first pyramidal level under consideration. Applying this methodology we group at the upper level street pixels belonging to wide, long roads (e.g. highways) and, using the more refined details of image, we arrive to smaller and shorter streets (see figure 3.1).

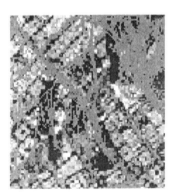 

Fig. 3.1. An example of street detection after fuzzy clustering. On the left the clustering results, on the right the corresponding ground truth.

4. Experimental Results

In this section we shall show how the theoretical considerations described previously are validated, using SAR data of the centre of Milan, Northern Italy and of Santa Monica, Los Angeles, CA.

For the Los Angeles data, we started by an FCM clustering in three classes, giving a classification rate (with threshold fixed to 0.5) of 76.3% and an index $i = 0.75$ (results can be seen for the street class in figure 4.1 left). Without classifying, and applying a PCM unsupervised clustering, initialized by the FCM partition matrix, the classification rate is increased to almost 85% and i to 0.84 (figure 4.1 right). Finally, all the pixels with a membership grade to the street class greater than 0.5 were used to extract the possible straight lines following the approach outlined in the precedent section. The street map obtained is the one shown in figure 3.1, together with the ground truth superimposed on the original image. Note that a further advantage of the fuzzy pyramidal GWHT is that now the street network is represented in vector format, with a large memory saving.

The same algorithm is shown in figure 4.2 applied to a small part of an IRS-1 panchromatic image of Milan (Northern Italy), with similar results; on the left we can see the FCM + PCM clustering results (classification rate 71.6%), and on the right the detected streets. Note, however, that since urban areas correspond to various different ways of life [8], the extraction of streets and roads as straight lines do not always correspond to the right action in different environments.

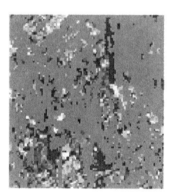 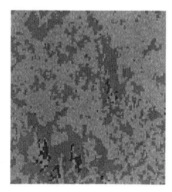

Fig. 4.1. The FCM (*left*) and PCM (*right*) classification results applied to an AIRSAR image of Santa Monica, Los Angeles, CA. Street class in FCM and street and building classes in PCM results are shown and the classification threshold is fixed to 0.5 in both cases.

Fig. 4.2. The FCM + PCM classification results of a small part of an IRS-1 panchromatic image of Milan, Northern Italy. Again, only three classes are considered and the classification threshold is fixed to 0.5 in both cases. On the right, the streets detected with the CWHT algorithm are shown.

5. Conclusions

This work presents some results of the application of a typical machine vision approach to the analysis of high resolution SAR images of urban environments. The proposed method is based on a *conceptually pyramidal* fuzzy clustering classification of the data, and the results are very encouraging. In the future, clustering modules based on fuzzy logic devoted to the extraction of urban structures other than streets need to be provided, as well as other street "followers" where straight roads can not be assumed.

Acknowledgement. The AIRSAR data of Los Angeles are ©JPL and were kindly provided by Dr. B. Houshmand.

References

1. R. N. Dave and R. Krishnapuram, "Robust clustering methods: a unified view," *IEEE Transactions on Fuzzy Systems*, vol. 5, no. 2, pp. 270–293, May 1997.
2. R. O. Duda and P. E. Hart, "Use of Hough Transformation to Detect Lines and Curves In Picture: Graphics and Image Processing," *Communications of the Association for Computing Machinery*, vol. 15, pp. 11–15, 1972.
3. M. A. Fischler, J. M. Tenenbaum, and H. C. Wolf, "Detection of roads and linear structures in low-resolution aerial imagery using a multisource knowledge integration technique," *Computer Graphics and Image Processing*, vol. 15, pp. 201–223, 1988.
4. P. Gong, D. J. Marceau and P. J. Howarth, "A comparison of spatial feature extraction algorithms for land-use classification with SPOT HRV data," *Remote Sensing of Environment*, vol. 40, pp. 137–151, 1992.
5. B. N. Haack, "Multisensor data analysis of urban environments," *Photogrammetric Engineering & Remote Sensing*, vol. 50, no. 10, pp. 1471–1477, 1984.

6. F. M. Henderson, and Z. G. Xia, "SAR applications in human settlement detection, population estimation and urban land use pattern analysis: a status report," *IEEE Transactions on Geoscience and Remote Sensing*, vol. 35, no. 1, pp. 79–85, January 1997.
7. F. M. Henderson, "An evaluation of SEASAT SAR imagery for urban analysis," *Remote Sensing of Environment*, vol. 12, pp. 439–461, 1982.
8. G. F. Hepner, B. Houshmand, I. Kulikov, and N. Bryant, "Investigation of the potential for the integration of AVIRIS and IFSAR for urban analysis," *Photogrammetric Engineering & Remote Sensing*, vol. 64, no. 8, pp. 813–820, 1998.
9. V. Lacroix and M. Acheroy, "Feature extraction using the constrained gradient," *Photogrammetry and Remote Sensing*, vol. 53, pp. 85–94, 1998.
10. A. Lopes, E. Nezry, R. Touzi and H. Laur, "Structure detection and statistical adaptive speckle filtering in SAR images," *International Journal of Remote Sensing*, vol. 14, no. 9, pp. 1735–1758, September 1993.
11. R. Samadan and J. F. Vesecky, "Finding curvilinear features in speckled images," *IEEE Transactions on Geoscience and Remote Sensing*, vol. 28, no. 4, pp. 66–673, July 1990.
12. U. Weidner and W. Forster, "Towards automatic building extraction from high resolution digital elevation models," *ISPRS Journal*, vol. 50, no. 4, pp. 38–49, 1995.
13. A. Winter, H. Maitre, N. Cambou and E. Legrand, "Entropy and multiscale analysis: a new feature extraction algorithm for aerial images," *Proceedings of International Conference on Acoustics, Speech and Signal Processing ICASSP'97*, vol. IV, pp. 2765–2768, April 21–24, Munich, Germany, 1997.
14. M. C. K. Yang, J.-S. Lee, C.-C. Lien and C.-L. Huang, "Hough transform modified by line connectivity and line thickness," *IEEE Transactions Pattern Analysis and Machine Intelligence*, vol. 19, no. 8, pp. 905–910, August 1997.
15. S. Y. K. Yuen, T. S. L. Lam, and N. K. D. Leung, "Connective Hough Transform," *Image and Vision Computing*, vol. 11, pp. 295–301, 1993.
16. Y. Zhang, "Information system for monitoring the urban environment based on satellite remote sensing: Shangai as an example," *Proceedings of International Geoscience and Remote Sensing Symposium, IGARSS'97*, vol. II, pp. 842–844, Singapore, August 1997.

Detecting Nets of Linear Structures in Satellite Images

Petia Radeva, Andres Solé, Antonio M. López, and Joan Serrat

Computer Vision Center and Departament d'Informàtica,
Universitat Autònoma de Barcelona, Edifici O, 08193–Cerdanyola, Spain.
e-mail: petia@cvc.uab.es

Summary. Recent developments in satellite sensors have made possible to analyse high resolution images with computer vision techniques. A large part of civil engineering objects appear as linear structures (e.g. channels, roads, bridges etc.). In this paper, we present a technique to detect linear features in satellite images based on an improved version of the level set extrinsic curvature. It allows the extraction of creases (ridge and valley lines) with a high degree of continuity along the center of elongated structures. However, due to its local nature, it can not cope with ambiguities originated by junctions, occlusion and branching of linear structures (e.g. hydrological or highway networks). To overcome this problem, we have applied a global segmentation technique based on geodesic snakes. It addresses line segmentation as a problem of detecting minimal length path. The geodesic snake looks for the path of minimal cost on a map that combines the information of the crease detector with the intensity of the original image. Some preliminary results on high resolution satellite images are presented.

1. Introduction

The image model underlying many image analysis methods considers images as the sampling of a smooth manifold, such as a graphic surface or landscape. Ridge/valley-like structures (lines in 2-d, lines and surfaces in 3-d, etc.) of a d-dimensional image, tend to be at the center of *anisotropic* grey-level shapes. These structures have been used in a number of applications as object descriptors, like the skull in Magnetic Resonance (MR) and Computed Tomography (CT) images, vessels in angiographies, oriented textures, handwritten characters, fringe patterns, fingerprints, aerial photographs depicting roads, rivers, cut-fires, etc. and, range and digital elevation model images.

The mathematical characterization of what is a ridge/valley-like structure is far from being a simple issue [11]. Several mathematical definitions have been proposed which we classify in three groups according to their scope:

- Creases: *local* descriptors that, basically, look for *anisotropic shapes*.
- Separatrices: *global* descriptors that result of linking the critical points of the image by means of slope lines, dividing the image in hydrologic hill/basin districts. Slopelines are curves which run orthogonally to the level curves, this is, they integrate the image gradient vector field.
- Drainage patterns: simulate the surface runoff induced in a landscape when raining. They have a *regional* or *semi-local* character.

In the literature we can still find several definitions within the crease class [5]. Among them, one of the most useful for its invariance properties is that based on level sets extrinsic curvature (LSEC). Given a function

$$L : \mathbf{R}^d \to \mathbf{R}$$

the level set for a constant c consists of the set of points

$$\{\mathbf{x}|L(\mathbf{x}) = c\}$$

Varying c we obtain all the level sets of L. The simplest situation is $d = 2$, when L can be thought as a topographic relief or landscape and the level sets are its level curves. It is well-known [8] that negative minima of the level curve curvature κ, level by level, form valley-like curves and positive maxima ridge-like curves (figure 1.1). They are characterized by the following local test, called the extremality criterion

$$e = \nabla \kappa \cdot \mathbf{v} = 0 \qquad (1.1)$$

where, $\nabla e \cdot \mathbf{v} < 0$, $\kappa > 0$ means ridge-like, v is the level curve tangent and $\nabla e \cdot \mathbf{v} > 0$, $\kappa < 0$ valley-like.

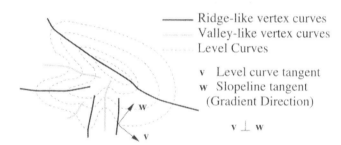

Fig. 1.1. Vertex curves: extrema of κ along the $\mathbf{v}$ direction, level by level.

The direct computation of extremality criteria as in equation (1.1) involves up to fourth order image derivatives combined into a complex expression (check for example [8] p.637 for the 2-d case). Nevertheless, the extrema of curvature of elongated structures in 2-d (vessels in angiography, fingerprint lines, roads in aerial images, range images etc.) are so high that we can circumvent the computational drawback by just computing κ as a creaseness measure and then apply a thresholding plus grouping step. Or just take the creaseness measure itself as a feature. This is by no means an infrequent situation for it is precisely for these anisotropic structures that ridge and valley lines are employed for description or detection.

However, the way LSEC is usually computed – directly from its equation – raises two problems. The first one is that it most probably results in an extremely large dynamic range of values, although there are only a few points with curvature at the upper and lower bounds, which we will call *outlayers*. The second is that it produces discontinuities: there appear gaps at places where we would not expect any reduction of creaseness because they are at the center of elongated objects and creaseness is a measure of medialness for gray-level elongated objects. Actually, these places are saddles and extrema along ridges and valleys.

To organize the set of image features in linear structures, different approaches have been considered. The most often-cited paper about road extraction is Fischler *et al.* [6]. The authors present a global formulation of the road problem minimizing a cost function on a graph defined over the pixel lattice. To cope with large images, Geman *et al.* [9] propose a decision tree approach based on the "divide-and-conquer" strategy applying a local statistic test to track linear image data. Another automated approach to find main roads in aerial images is presented by Barzohar *et al.* [1] where geometrical-probabilistic models for road image generation are constructed based on local analysis of the image. By estimating the maximum a posterior probability of the image windows they obtain an optimal global estimate of the road present. Fua [7] uses a snake-based optimization technique to reconstruct the relief from stereo images and extract drainage patterns. Cohen and Kimmel [4] present an optimization approach based on finding geodesics on a surface that can be done in near real time on a regular workstation using recent dynamic programming implementations.

In section 2 we review the traditional way of computing LSEC, analyze the above mentioned problems and propose an alternative way that avoids them. Section 3 presents a further enhancement which we call the structure tensor-based LSEC and show the results of our approach applied to the problem of linear feature extraction in satellite images. In section 4 we discuss the problem of detecting linear structures as a problem of finding a minimal path in images. Different applications in aerial images are outlined in section 5. Conclusions are summarized in section 6.

2. LSEC Based on the Gradient Vector Field

The level sets extrinsic curvatures in Cartesian coordinates for the two-dimensional cases is:

$$\kappa = (2L_x L_y L_{xy} - L_y^2 L_{xx} - L_x^2 L_{yy})(L_x^2 + L_y^2)^{-\frac{3}{2}} \quad (2.1)$$

where L_α denotes the first order partial derivative of L with respect to α, $L_{\alpha\beta}$ the second order partial derivative with respect to α and β. To obtain derivatives of a discrete image L, in a well-posed manner, we have approximated image derivatives by finite differences of the Gaussian smoothed image, instead of convolution with Gaussian derivatives, because it is computationally more efficient and produces very similar results. Therefore, our differentiation scheme is:

$$L_\alpha(\mathbf{x}; \sigma_D) = \frac{\partial}{\partial \alpha}(L(\mathbf{x}) * G(\mathbf{x}; \sigma_D)) \approx \delta_\alpha L(\mathbf{x}; \sigma_D) \qquad (2.2)$$

where δ_α represents a partial finite difference along the discrete α axis and σ_D is the standard deviation of the Gaussian kernel and the differentiation scale. Centered finite differences are used because of their invariance under rigid movements of the parametric plane of the image and their high regularizing power.

In the 2-d case, if we travel along the center of folded anisotropic structures we go up and down on the relief, passing through non-degenerated and isolated maxima, saddles and minima. We have found that the computation of κ according to equation (2.1), produces gaps and outlayers on the creaseness measure around this type of critical points (e.g. figure 2.1), as well as on the center of elongated grey-level objects having a short dynamic range along this center. The gaps and outlayers are produced independently of the scheme used to derive the derivatives. Moreover, the problem is not due to a low gradient magnitude, since actual critical points are placed at interpixel coordinates, and gaps appear even when gradient is far away from the zero of the machine, that is, at pixels where κ is well-defined according to equation (2.1).

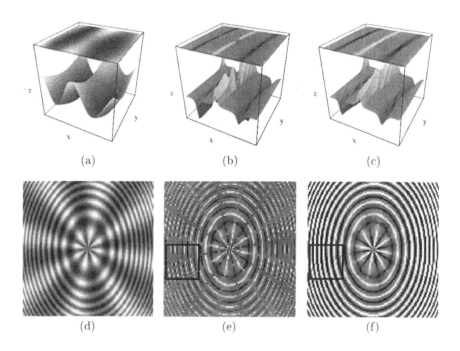

Fig. 2.1. (a) Landscape with regions having valley-like and a ridge-like aspect, (d) synthetic image, (b) and (e) κ, (c) and (f) $\bar{\kappa}$.

To avoid those problems we present an alternative way of computing creaseness from the level sets extrinsic curvature. In vector calculus there is a well-known operator called divergence which measures the degree of parallelism of a vector field and, therefore, of its integral lines. For a d-dimensional vector field $\mathbf{W} : \mathbf{R}^d \to \mathbf{R}^d, \mathbf{W}(\mathbf{x}) = (W^1(\mathbf{x}), ..., W^d(\mathbf{x}))$, its divergence is defined as:

$$\mathrm{div}(\mathbf{W}) = \sum_{i=1}^{d} \frac{\partial W^i}{\partial x^i} \qquad (2.3)$$

Now, if we denote by $\mathbf{0}_d$ the d-dimensional zero vector, then we can define $\bar{\mathbf{w}}$, the normalized gradient vector field of $L : \mathbf{R}^d \to \mathbf{R}$, as:

$$\bar{\mathbf{w}} = \begin{cases} \mathbf{w}/\|\mathbf{w}\| & \text{if } \|\mathbf{w}\| > 0 \\ \mathbf{0}_d & \text{if } \|\mathbf{w}\| = 0 \end{cases} \qquad (2.4)$$

It can be shown that:

$$\kappa_d = -\mathrm{div}(\bar{\mathbf{w}}) \qquad (2.5)$$

Equations (2.1) and (2.5) are of course equivalent. However, the same derivatives approximation of equations (2.1) and (2.5) gives different results, namely, the discretization of (2.5) avoids the gaps and outlayers that equation (2.1) produces on creases. Thus, from now on, we denote by κ and $\bar{\kappa}$ the discrete versions of the level sets extrinsic curvature for $d = 2$, according to equations (2.1) and (2.5), respectively. For higher values of d, κ_d and $\bar{\kappa}_d$ will be used. Figure 2.1 compares the behavior of κ and $\bar{\kappa}$. Notice how gaps disappear in figure 2.1(f). Furthermore, it can be shown that if we approximate the derivatives of a d-dimensional image with centered differences then $|\bar{\kappa}_d| \leq d$, reaching $-d$ at minima and d at maxima.

3. LSEC Based on the Structure Tensor

Once we have established $\bar{\kappa}$ as a good creaseness measure, we can go further and enhance it by modifying the gradient vector field of the image before applying the divergence operator. Given the usual configurations of the gradient vector field around a crease line displayed in figure 3.1(a), we want to filter the gradient vector field in order to increase the degree of collapse/repulsion as shown in figure 3.1(b) and consequently its creaseness. At the same time, regions where there is neither a clear collapsing or repulsion must not change. This filtering can be carried out in a natural way through the so-called *structure tensor*, which is a well-known tool for analyzing oriented textures [2].

Let us assume that, at any given point $\mathbf{x} \in \mathbf{N}^d$ there is a single dominant orientation at most, taking into account a weighted neighbourhood of size σ_I around it, $\mathcal{W}(\mathbf{x}; \sigma_I)$. The structure tensor is represented by the symmetric and semi-positive definite $d \times d$ matrix:

$$\mathbf{M}(\mathbf{x}; \sigma_I; \sigma_D) = \mathcal{W}(\mathbf{x}; \sigma_I) * (\mathbf{w}(\mathbf{x}; \sigma_D)\mathbf{w}(\mathbf{x}; \sigma_D)^T) \qquad (3.1)$$

Detecting Nets of Linear Structures in Satellite Images   309

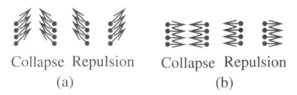

Fig. 3.1. Collapse and repulsion of vectors in (b) is higher than in (a).

where the convolution is element-wise. A suitable choice for the window is a Gaussian, i.e. $\mathcal{W}(\mathbf{x};\sigma_I) = G(\mathbf{x};\sigma_I)$. The dominant gradient orientation is given by the eigenvector which corresponds to the highest eigenvalue of $\mathbf{M}$. We can enhance creaseness by first performing the eigenvalue analysis of the structure tensor $\mathbf{M}(\mathbf{x};\sigma_I;\sigma_D)$. The normalized eigenvector $\tilde{\mathbf{w}}(\mathbf{x};\sigma_I;\sigma_D)$, which corresponds to the highest eigenvalue of $\mathbf{M}(\mathbf{x};\sigma_I;\sigma_D)$, gives the predominant gradient orientation at $\mathbf{x}$. Then, the divergence operator is applied to the $\tilde{\mathbf{w}}$ vector field, giving rise to a new creaseness measure:

$$\tilde{\kappa}_d = -\mathrm{div}(\tilde{\mathbf{w}}) \qquad (3.2)$$

Given the objective to extract linear structures in satellite images, we preprocess the image applying the structure tensor. Figure 3.2(a) shows a high-resolution satellite image and its filtered image by the structure tensor (figure 3.2(b)). The predominant orientation of image structures can be

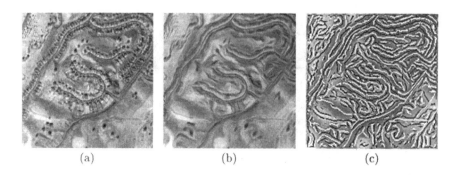

Fig. 3.2. (a) Original satellite image, (b) filtered image using the structure tensor, (c) extracted ridges (in black) and valley points (in white).

noted that is represented by different linear structures as roads or building alignments. Structure tensor allows to improve the detection of valleys and ridges (figure 3.2(c)) in the image. A comparison between the preprocessing by a Gaussian filter and the structure tensor is given in figure 3.3(a), (b) and (c). It is clear that both preprocessing techniques have opposite effects. Figure 3.4 shows the result of depicting man-made linear structures after

310 P. Radeva et al.

applying the structure tensor operator on several high resolution satellite images. Figure 3.5 compares it with other drainage pattern operators.

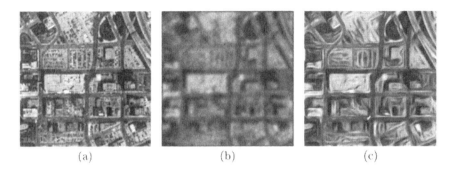

Fig. 3.3. (a) Original aerial image, (b) preprocessing by a Gaussian filter, (c) preprocessing by the structure tensor.

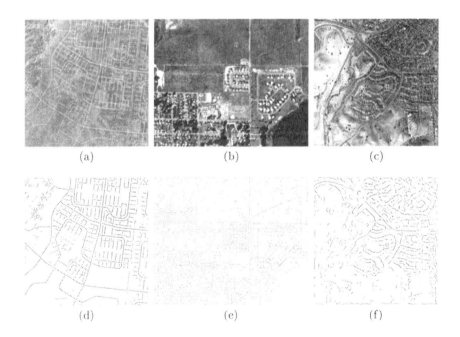

Fig. 3.4. Satellite images (a,b,c) and the corresponding curvature map $\tilde{\kappa}$ (d,e,f).

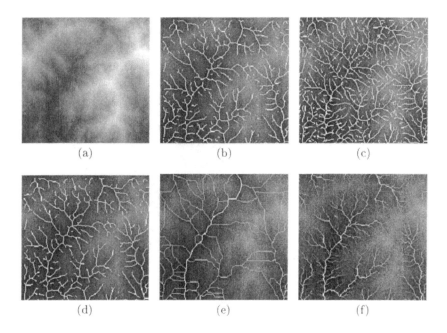

Fig. 3.5. (a) Digital elevation model, (b) κ, (c) $\nabla \kappa \cdot v = 0$, (d) $\tilde{\kappa}$, (e) watershed, (f) algorithm of Soille and Gratin [13].

4. Detecting Linear Structures

The problem of finding roads, rivers, and other linear structures highly depends on the robustness of the image feature extraction. At the same time, high-level image processing is still necessary in order to select and organize image features in semantical units. Recently, active contour models have become a standard energy-minimization technique to organize image features. These take into account different constraints sush as smoothness, continuity [10], closeness and guide by an approximate model [12], etc. These constraints regularize the problem of image feature organization leading to a unique solution as a function of the initial conditions.

Although linear structures may be characterized by constraints of smoothness, continuity, etc. they are not sufficient to avoid that the snake is too dependent of its initialization. Following an iterative procedure to minimize its energy, the snake can not avoid the local minima i.e. to be attracted by false image features. To cope with this shortcoming, Cohen and Kimmel propose in [4] a new approach to find the global minimum of energy by minimizing curves based on classic active contour models [10] and geodesic snakes [3]. The advantage of this approach is that, given two end points of a curve, it is assured that the snake will find the global minimum of the energy. The main idea is that the minimization of the classical energy of the snakes is posed as

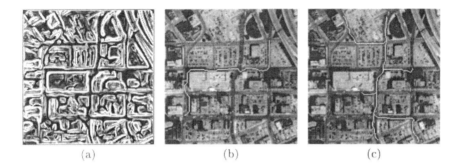

(a)   (b)   (c)

Fig. 4.1. (a) extracted contours (in white) and valleys (in black), (b) finding path of minimal length between two points, (c) finding different paths of minimal length to a given point (derived from the images of figure 3.3(a), (b) and (c)).

a problem of finding local geodesics in a Riemannian metric computed from the image.

Linear structures usually appear in satellite images as bright or dark smooth (but not necessarily straight) lines. Therefore, looking for linear structures on the ridge and valley map of the original image provides more robust results than looking for high or low intensity pixels (as in [4]). We extract ridge and valley points (see figure 3.2(c)) and generate a potential which takes lower values in the image features of interest. For example, roads and rivers are characterized as valley lines (see figure 3.2(c)). Additionally, one can find the contours of the image, usually corresponding to building borders, as well as crest points, appearing close to valley points (see figure 4.1(a)). High values are assigned at this location to the potential serving as a barrier to avoid cases where the snake is attracted by false data in the image. Note that in this case the potential should not be a continuous surface compared to the classical snakes.

The problem of finding linear structures is posed as a problem of finding as many as possible image features of interest in as short as possible path between two given end points. The 'best fit' question leads to algorithms seeking for the minimal path, i.e. paths along which the integration over the potential P is minimal. The snake energy is determined as follows [4]:

$$\int_\Omega \omega ||C_s||^2 + P(C) ds = \int_\Omega \tilde{P}(C) ds$$

where C is the snake curve, s is its arc-length, ω is a weight parameter and $\tilde{P}(p) = \omega + P(p)$. The energy is minimized on the space of all curves connecting two given end points: $C(0) = p_0$ and $C(L) = p_1$, where L is the curve length.

The snake with minimal energy is found in two steps [4]: first, a surface of minimal action U_0 is constructed, where at each point p of the image plane, a value of the surface U_0 is assigned equal to the minimal energy integrated along a path starting at p_0 and ending at p: $U_0(p) = \inf_{C(L)=p} \int_C \tilde{P} ds$. Let us consider the set of equal energy contours L in 'time' where t is the value of the energy as level sets of the surface of minimal action. Starting at the initial point p_0, the level sets can be obtained by the following evolution equation: $\frac{\delta L(v,t)}{\delta t} = \frac{1}{\tilde{P}} \mathbf{n}(v,t)$, where $\mathbf{n}(v,t)$ is the normal to the closed curve L. Once we construct the surface of minimal action, the algorithm assures that the surface is convex with a minimum at the point p_0 and thus, starting from any point p_1 and following the gradient descent of the surface, we can find the shortest path between p_0 and p_1.

5. Application of Minimal Path Algorithm to Satellite Images

The algorithm of minimal geodesic path has several advantages for the detection of linear structures in satellite images. The fact that it finds the shortest path between two points, optimizing a certain criterion (e.g. to visit the maximum number of valley points), allows to detect lines even when they are not completely connected. This is very useful because no algorithm for extraction of crease points can assure the detection of continuous lines. Another advantage is that points of the structures are not supposed to lie on a straight line and no piece-wise linear assumption is made. At the same time, the algorithm allows to find short and smooth linear structures. In figure 5.1(c) an example of finding roads with curvilinear shape is illustrated. The image features used to construct the potential in this case are the valley and ridge points. Figures 5.1(a) and (b) show two surfaces of minimal action where the darkest pixel corresponds to the initial point. The paths are obtained following the gradient descent of the surface beginning from different points selected by the user.

Another example to find the shortest path between two points in a urban scene is given in figure 4.1(b). This task is very frequent in navigation applications.

Sometimes, it is of interest to find more than one path to the same point starting from different positions (e.g. different rivers going to the same delta, or detecting possible points of peak traffic in road nets). To detect tree-shaped linear structures we construct the surface of minimal action with a minimum at the root of the tree and go down from the leaves to reach different paths. An example of finding the shortest paths to the same point is given in figure 4.1(c).

As mentioned above, the algorithm works even in case of discontinuous sets of data. Such situations occur in case of occlusions of the roads (e.g. road intersections). In figure 5.2 a case of finding the road in the case of a highway crossing is shown. The road is extracted as a crease line creating the

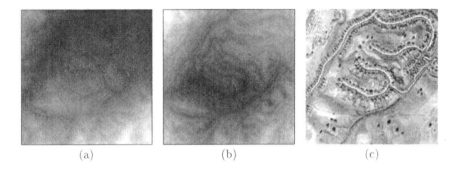

Fig. 5.1. (a) and (b) two surfaces of minimal action, (c) detected curvilinear paths (derived from the images of figure 3.2(a), (b) and (c)).

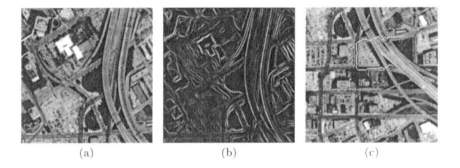

Fig. 5.2. (a) Original aerial image of highway scene, (b) extracted contours, (c) finding path of minimal length between two points in case of road occlusion.

potential with lower values in crease data and higher values in contour data serving as barriers to restrict the path to follow the road.

6. Conclusion

Level sets extrinsic curvature is a creaseness measure achieving several desirable properties that acts as medialness feature for grey-level objects. In this paper we have first pointed out the problem of outlayers and gaps when computing it directly. We have proposed an alternative way of computing LSEC that overcomes the above mentioned problems. We have also proposed a modified version adapting the structure tensor filtering, designed for oriented texture analysis, to improve its results.

To detect linear structures in satellite images we have appled the algorithm for finding optimal geodesics on the crease map extracted from the image. We have showed that this approach allows us to detect linear features (e.g. roads) from a disconnected set of data that form smooth but not necessarily straight shapes. The detection of optimal paths in urban scenes as well as the identification of linear networks even in the case of junctions, occlusions and branching can be of special interest for urban planning and control.

Acknowledgement. This research has been partially funded by CICYT projects TIC97-1134-C02-02 and TAP96-0629-C04-03.

References

1. M. Barzohar and D. Cooper, "Automatic finding of main roads in aerial images by using geometric-stochastic models and estimation", *IEEE Transactions on Pattern Analysis and Machine Intelligence*, vol. 18, no. 7, 1996.
2. J. Bigun, G. Granlund, and J. Wiklund, "Multidimensional orientation estimation with applications to texture analysis and optical flow", *IEEE Transactions on Pattern Analysis and Machine Intelligence*, vol. 13, no. 8, pp. 775–790, 1991.
3. V. Caselles, R. Kimmel and G. Sapiro, "Geodesic active contours", in *Proceedings International Conference on Computer Vision (ICCV'95)*, Cambridge, USA, pp. 694–699, 1995.
4. L. D. Cohen and R. Kimmel, "Global minimum for active contour models: A minimal path approach", in *Proceedings IEEE International Conference on Computer Vision and Pattern Recognition (CVPR'96)*, San Francisco, USA, pp. 666–673, 1996.
5. D. Eberly, R. Gardner, B. Morse, S. Pizer, and C. Scharlach, "Ridges for image analysis", *Journal of Mathematical Imaging and Vision*, vol. 4, pp. 353–373, 1994.
6. M. A. Fischler, J. M. Tenenbaum and N. C. Wolf, "Detection of roads and linear structures in low-resolution aerial imagery using a multisource knowledge integration technique", *Computer Vision, Graphics, and Image Processing*, vol. 15, no. 3, pp. 201–223, 1981.
7. P. Fua, "Fast, accurate and consistent modeling of drainage and surrounding terrain", *International Journal of Computer Vision*, vol. 26, no. 3, pp. 215–234, 1998.
8. J. M. Gauch and S. M. Pizer, "Multiresolution analysis of ridges and valleys in grey-scale images", *IEEE Transactions on Pattern Analysis and Machine Intelligence*, vol. 15, pp. 635–646, 1993.
9. D. Geman and B. Jedynak, "An active testing model for tracking roads and satellite images", *IEEE Transactions on Pattern Analysis and Machine Intelligence*, vol. 18, no. 1, pp. 1–14, 1996.
10. M. Kass, A. Witkin and D. Terzopoulos, "Snakes: active contour models", in *Proceedings of International Conference on Computer Vision (ICCV'87)*, London, pp. 259–268, 1987.
11. A. M. López and J. Serrat, "Tracing crease curves by solving a system of differential equations", in B. Buxton and R. Cipolla, eds., *Proceedings 4th European Conference on Computer Vision*, vol. 1064 of *LNCS*, pp. 241–250, Springer-Verlag, 1996.
12. P. Radeva, J. Serrat and E. Martí, "A snake for model-based segmentation", *International Conference on Computer Vision (ICCV'95)*, MIT, USA, pp. 816–821, June 1995.

13. P. Soille and C. Gratin, "An efficient algorithm for drainage network extraction on DEMs", *Journal of Visual Communication and Image Representation*, vol. 5, pp. 181–189, 1994.

Satellite Image Segmentation Through Rotational Invariant Feature Eigenvector Projection

Albert Pujol, Andrés Solé, Daniel Ponsa, Javier Varona, and Juan José Villanueva

Computer Vision Center & Departament d' Informàtica
Universitat Autònoma de Barcelona, 08193 Bellaterra (Barcelona), Spain.
e-mail: {albert,andres,dani,xaviv,juanjo}@cvc.uab.es

Summary. In this chapter a method for rotational invariant object recognition is developed. The proposed method is based on the principal component analysis technique applied to the set of images formed when a target image is rotated, the theoretical results of this approach permit us to automatically design a set of image filters that, when properly combined, serve as a rotational invariant shape discriminant tool. This method is applied to satellite image shape-based segmentation.

1. Introduction

There are many techniques for recognizing objects. The problem arises when the objects to be recognized are not in a pre-determined position, orientation, or scale. In these cases it is necessary to use an invariant object recognition method. These methods maintain the system performance independently of a particular set of image transformations.

The principal approach to invariant object recognition consists of the extraction of image information that remains constant in spite of the image transformation applied. The extracted information can be an image transformation or a set of features that become an invariant signature of the object. Reviews on invariant recognition techniques can be found in [6, 13, 14]. Some widely used invariant image transformations and signatures are moment based invariants [1], integral measures such as density chord estimation or Hough methods. Fourier or Fourier-Mellin transforms, and image contours differential geometry invariants. Once an invariant signature is obtained a pattern classification method such as neural networks, Principal Components Analysis, or Bayesian classifiers can be applied [12].

Principal component analysis (PCA) [2] also referred as Hotelling or Karhunen-Loève transform is a widely applied statistical technique for data dimensionality reduction. This feature makes PCA a truly efficient tool widely used in classification methods. A lot of its applications such as face recognition [5] can be found in the object recognition field. Nonetheless PCA is too much sensitive to image transformations. Some approaches have been developed to give invariant capabilities to the classical PCA methods; thus

Picard et al. [8] proposes using PCA on the Fourier modulus of the input vectors. This two step approximation however, causes some loss of information, because the phase information is discarded.

We propose a 2D invariant object recognition method related to PCA and Fourier transform that preserves its efficiency while at the same time offers an invariant classification behaviour.

This chapter gives a description of the proposed method and its application to a rotational invariant airplane detection. The obtained results are discussed and future developments are outlined.

2. Rotational Invariant Eigenspaces

The goal of the object recognition methods is to classify an image into the appropriate model classes. It is desirable that a recognition method classifies an image independently of its position, orientation and scale. Therefore, the method should be invariant to similarity transformations.

If we represent the objects as $M \times N$ pixel images, each one could be considered a point in the $\mathbf{R}^{M \times N}$ space. The rotation transformations of the model image result in other different points in $\mathbf{R}^{M \times N}$. This set of $\mathbf{R}^{M \times N}$ is the rotations orbit of the image.

In [7] a method for 3D invariant object recognition and pose estimation is presented. The pose estimation is based on the determination of the space spanned by images of an object taken at different poses and illuminations. This space is compressed using the Karhunen-Loève transformation and is called the object eigenspace.

The proposed method exploits the fact, previously used in [3], that given an image object, the covariance matrix, Σ, of all its 2D rotations can be determined easily from only one image. Once the covariance matrix is constructed, we can determine a subspace that is invariant to the considered transformation. Furthermore, it is easy to compute the Euclidean distance of any object to these subspaces. Finally, dimensionality reduction techniques are employed to reduce the classification task computational cost. Unlike the work presented in [3], we will see that the basis vectors that define the rotations orbit of an image can be determined directly from its Fourier transform.

In the next section, we show how to construct a covariance matrix that represents a subspace invariant to rotational transformations. A similar approach could be used in any transformation expressed as cyclic translation.

3. Computing the Rotation's Orbit Eigenvectors and Eigenvalues

Our approach is based on the definition of a covariance matrix that represents an image model and all its possible rotations. For the matrix construction

Satellite Image Segmentation   319

we re-map the image, $I(x,y)$, from Cartesian coordinates (x,y) to the polar coordinate system (r,θ).

The reason to re-map the raw onto polar coordinates is that a rotation in raw coordinates becomes a translation in polar coordinates. This is an interesting property that simplifies the treatment of the invariance under rotation. First, to simplify the explanation of the covariance matrix construction, we consider the one dimensional case.

3.1 Covariance Matrix for the One Dimensional Case

If we consider a discrete ring in the (x,y)-plane centred at the origin, where each of its cells has assigned a different grey level, its projection in polar coordinates becomes a discrete vertical segment. Then, if we consider a central rotation of angle $d\theta$, its projection is equal to the translation of the non-rotated projection $d\theta$ in the θ direction (figure 3.1).

Our objective is to construct a covariance matrix that takes advantage of this behaviour. Therefore, we consider into the covariance matrix all the possible rotations of the original ring. Thus, we can generate directly a matrix that takes into account how the rotations are mapped in the polar space. In fact, if the non rotated ring projection is a discrete vector, $\mathbf{x}$, a rotation by k discrete units of the original data results in the following cyclical translation in the polar space:

$$\mathbf{x}^T = (x_{n-k}, \ldots, x_n, x_0, x_1, \ldots, x_{n-1-k}) \tag{3.1}$$

We note the k positions translated vector as,

$$\mathbf{x}^{T(k)} = (x_{(0+k)\bmod n}, x_{(1+k)\bmod n}, \ldots, x_{(n-1+k)\bmod n}) \tag{3.2}$$

i.e.,

$$\begin{aligned}\mathbf{x}^{T(0)} &= (x_0, x_1, \ldots, x_{n-1}) \\ \mathbf{x}^{T(1)} &= (x_{n-1}, x_0, \ldots, x_{n-2}) \\ \mathbf{x}^{T(2)} &= (x_{n-2}, x_{n-1}, \ldots, x_{n-3}) \\ &\vdots \\ \mathbf{x}^{T(n-1)} &= (x_1, x_2, \ldots, x_0)\end{aligned} \tag{3.3}$$

Consequently, the mean vector, $\boldsymbol{\mu}$, of all the possible rotations has the same value in all positions because the same data is considered in each calculation, but in different order:

$$\boldsymbol{\mu} = E[x^{(k)}] = \frac{1}{n}\sum_{i=0}^{n-1} x^{(i)} \tag{3.4}$$

where $\mathbf{x}_{(i)}$ is the result of applying a rotation of i discrete units to the original vector. Thus, by equation (3.2), the components of the $\boldsymbol{\mu}$ vector are:

$$\mu_i = \frac{1}{n}\sum_{j=0}^{n-1} x_j^0 \quad \forall\, i = 0 \cdots n-1 \tag{3.5}$$

Next, we define the covariance matrix, Σ of all the possible rotations:

$$\Sigma = E[(\mathbf{x} - \mu)(\mathbf{x} - \mu)^T] \tag{3.6}$$

The original vector, $\mathbf{x}$, is normalized, in order to simplify the calculations, by subtracting the mean vector,

$$\mathbf{x}' = \mathbf{x} - \mu \tag{3.7}$$

then the covariance matrix will be,

$$\Sigma = E[\mathbf{x}'\mathbf{x}'^T] \tag{3.8}$$

with elements,

$$\Sigma_{i,j} = \frac{1}{n}\sum_{k=0}^{n-1}(x_i^{'(k)} x_j^{'(k)}) \tag{3.9}$$

where n is the number of vectors, i.e., the original vector and its n-1 cyclic translations. Rewriting the covariance matrix using Equation 3.2:

$$\Sigma_{i,j} = \frac{1}{n}\sum_{k=0}^{n-1}(x'_{(i+k)\bmod n} x'_{(j+k)\bmod n}) = \frac{1}{n}\sum_{k=0}^{n-1}(x'_k x'_{(j-i+k)\bmod n}) \tag{3.10}$$

$\Sigma_{i,j}$ is equal to the sum of products of the couples of elements of the input vector separated for $(j-i) \bmod n$ elements:

$$\Sigma_{i,j} = \Sigma_{k,l} \quad \text{if} \quad (j-i) \bmod n = (l-k) \bmod n \tag{3.11}$$

or,

$$\Sigma_{i,j} = \Sigma_{(i+t)\bmod n, (j+t)\bmod n} \quad \forall\, t \in N \tag{3.12}$$

Thus only the first row of the covariance matrix should be computed. It can be seen in equation (3.10) that the vector defined for the elements of the first row of the covariance matrix are the autocorrelation of the normalized vector $\mathbf{x}'$. And each row is equal to its upper one but cyclicly shifted one position. Thus the resultant covariance matrix is Toeplitz and symmetric.

For example, if we have a one dimensional vector of size $N = 6$, then the covariance matrix has the following form:

$$\Sigma = \begin{pmatrix} \Sigma_0 & \Sigma_1 & \Sigma_2 & \Sigma_3 & \Sigma_2 & \Sigma_1 \\ \Sigma_1 & \Sigma_0 & \Sigma_1 & \Sigma_2 & \Sigma_3 & \Sigma_2 \\ \Sigma_2 & \Sigma_1 & \Sigma_0 & \Sigma_1 & \Sigma_2 & \Sigma_3 \\ \Sigma_3 & \Sigma_2 & \Sigma_1 & \Sigma_0 & \Sigma_1 & \Sigma_2 \\ \Sigma_2 & \Sigma_3 & \Sigma_2 & \Sigma_1 & \Sigma_0 & \Sigma_1 \\ \Sigma_1 & \Sigma_2 & \Sigma_3 & \Sigma_2 & \Sigma_1 & \Sigma_0 \end{pmatrix}$$

where $(\Sigma_0, \Sigma_1, \Sigma_2, \Sigma_3, \Sigma_2, \Sigma_1)$ is the autocorrelation of the normalized vector $\mathbf{x}'$.

The basis vectors of the compressed subspace spanned for the rotation's orbit are the eigenvectors of the covariance matrix with lower eigenvalues. It can be seen [11] that the eigenvalues of a Toeplitz matrix are the Fourier transform of its first row elements. It can be seen also that its eigenvectors are the basis vectors for the discrete Fourier transform associated to it. The Fourier transform of the autocorrelation of a vector is the square of the modulus of the Fourier transform of the original vector [9]. Thus, the eigenvalues associated to each eigenvector are real valued and can be computed directly from the Fourier transform of the normalized vector. Since the eigenvalues of the covariance matrix have only a real part, its eigenvectors will be the real part of the Fourier functions. That is, each eigenvector will be a cosine at a different frequency.

Thus, the kth eigenvector will be:

$$\mathbf{u}_k = (u_{k,0}, u_{k,1}, \ldots, u_{k,n-1}) \quad where \quad u_{k,i} = \cos(\frac{2\pi k}{n}i) \quad \forall\, k \in [0, n-1] \tag{3.13}$$

and its eigenvalue:

$$\lambda_k = [F(\mathbf{x}')]_k \quad \forall\, k \in [0, n-1] \tag{3.14}$$

3.2 The Two Dimensional Case

Consider now M concentric rings instead of the single one used in the previous section. When we take the intensity values of the image under these rings, (see figure 3.1) we have an $M \times N$ 2D tensor. When a rotation on the original image is applied, each of the M columns are transformed with the same cyclical translation. Thus the resulting 2D tensor of rotating the original image by k discrete units will be :

$$\mathbf{x} = \begin{pmatrix} x_{0,(0+k)\bmod n} & x_{1,(0+k)\bmod n} & x_{2,(0+k)\bmod n} & \cdots & x_{n-1,(0+k)\bmod n} \\ x_{0,(1+k)\bmod n} & x_{1,(1+k)\bmod n} & x_{2,(1+k)\bmod n} & \cdots & x_{n-1,(1+k)\bmod n} \\ \vdots & \vdots & \vdots & \ddots & \vdots \\ x_{0,(n-1+k)\bmod n} & x_{1,(n-1+k)\bmod n} & x_{2,(n-1+k)\bmod n} & \cdots & x_{n-1,(n-1+k)\bmod n} \end{pmatrix} \tag{3.15}$$

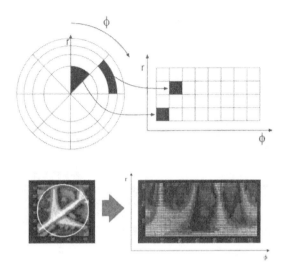

Fig. 3.1. Projection into polar coordinates (2D). The wideness of each ring is taken so that each ring area remains equal.

We can repeat the covariance matrix calculation like in the 1D case. Now, to determine the covariance matrix we need the 2D tensors of each model, and its $n-1$ possible cyclic translations. Therefore, the 2D tensors set from which we will construct the covariance matrix is $\{\mathbf{x}^{(0)}, \mathbf{x}^{(1)}, \ldots, \mathbf{x}^{(k)}, \ldots, \mathbf{x}^{(n-1)}\}$.

Unlike the case of 1D vectors, in which the mean vector components have the same value, we now have a mean tensor with M different components, one for each column (ring in the original image):

$$\boldsymbol{\mu} = \begin{pmatrix} \mu_0 & \mu_1 & \mu_2 & \cdots & \mu_{n-1} \\ \mu_0 & \mu_1 & \mu_2 & \cdots & \mu_{n-1} \\ \mu_0 & \mu_1 & \mu_2 & \cdots & \mu_{n-1} \\ \vdots & \vdots & \vdots & \ddots & \vdots \\ \mu_0 & \mu_1 & \mu_2 & \cdots & \mu_{n-1} \end{pmatrix} \quad (3.16)$$

where

$$\mu_i = \frac{1}{n} \sum_{j=0}^{n-1} x_{i,j} \quad (3.17)$$

Then, the input tensor is normalized subtracting the mean, and is used to compute the covariances. Now, we have a matrix of $(M \times N) \times (M \times N)$ covariance coefficients. Thus, the covariance between a tensor element (i,j) and the element (k,l) will be:

$$\Sigma_{i*N+j,k*N+l} = \frac{1}{n}\sum_{t=0}^{N-1}\left(x'_{i,(j+t)\bmod n}x'_{k,(l+t)\bmod n}\right)$$
$$= \left(x'_{i,t}x'_{k,(l-j+t)\bmod n}\right) \quad (3.18)$$

The 1D case can be extended to 2D images in polar coordinates and it can be seen, that to construct the eigenvalues of the rotations orbit of a 2D image **x**, the following algorithm can be applied:

1. Transform the original image **x** into polar coordinates
2. Compute the mean of each row. Subtract it from the row elements.

$$\mathbf{x}'_{i,j} = \mathbf{x}_{i,j} - \mu_i \quad \text{where} \quad \mu_i = \frac{1}{n}\sum_{j=0}^{N-1} x_{i,j} \quad \forall\, i = 0\cdots M-1 \quad (3.19)$$

3. Compute the square of the modulus of the FFT of each row.

$$\mathbf{f_i} = \left|\mathcal{F}(\mathbf{x}'_i)\right|^2 = (f_{i,0}, f_{i,1}, ..., f_{i,N-1}) \quad (3.20)$$

4. The N eigenvalues will be the sum of the coefficients determined in 3.20.

$$\lambda_j = \sum_{k=0}^{M-1} f_{k,j} \quad \forall\, j = 0\cdots N-1 \quad (3.21)$$

The eigenvector associated to the kth eigenvalue is an image composed of concentric rings. The grey levels inside each ring vary sinusoidally with respect to the angle. So each ring depicts a cosine at frequency $\frac{2\pi k}{M}$. All the cosines of the rings of an eigenvector are tuned to the same frequency and the rings only differ by their phase and magnitude. The phase and magnitude are calculated through the Fourier transform of the ring's elements.

1. Construct the element j of each row i of the kth eigenvector as:

$$\mathbf{u}_{i,j} = m_i \cos(\frac{2\pi w_k}{M}j + \phi_i) \quad (3.22)$$

where m_i is the magnitude of the Fourier transform component of frequency w_k of the row i and ϕ_i is its phase.
2. Re-map the eigenvector from polar to Cartesian coordinates.
3. Normalize the whole eigenvector.

Step 1 and 2 can be merged, so that instead of scanning the polar coordinates image and re-map to Cartesian coordinates, it can be directly done scanning the Cartesian coordinates image, and computing equation (3.22) for the radius and phase that corresponds to each pixel.

324 A. Pujol et al.

4. Experimental Results: Fast Rotational Invariant Airplane Detection in Satellite Images

Given an image **y**, it is classified with its nearest subspace. The distance from the image to each subspace is computed as the difference between **y** and the vector reconstructed from the **y** projection in the considered subspace. We have used this approach successfully in a previous work [10] for rotational invariant coin shape identification. In this work the images of coins to be classified were centred before computing its subspace i.e. the coin position was normalized. Instead of this, now we want to identify an object in whichever position and orientation within the image, without position normalization. That is, we will use the method for rotational invariant pattern matching and as an example we will show its capabilities in localizing airplanes in satellite images.

The goal of this experiment is to localize the airplanes of figure 4.1. To construct the subspace of the airplane and its rotations we have taken the image of the airplane marked with a white square in the same figure. The target airplane image of 64 × 64 pixels is resampled to polar co-ordinates (figure 3.1), then the eigenvectors construction method is applied. The sorted magnitudes of the eigenvalues computed with the proposed method can be seen in figure 4.2 where the 8 highest eigenvalues have been taken. The eigenvectors of the selected eigenvalues have been reconstructed (figure 4.3). These 8 eigenvectors are the basis of the rotations orbit eigenspace of the model airplane.

If all 64 × 64 image windows have equal energy levels then, it can be shown [4] that when the distance from a window to the subspace is minimized the magnitude of the projection over the subspace is maximized. Thus instead of computing the distance of each subwindow to the rotation's orbit airplane

Fig. 4.1. Original image and the result of filtering with the first 8 eigenvectors.

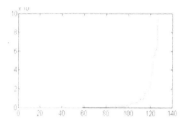

Fig. 4.2. Sorted eigenvalues and airplane rotation's orbit mean image.

Fig. 4.3. Principal eigenvectors of the model airplane rotations orbit.

subspace, the magnitude of its projection is computed, and then an image window is of the airplane class if the magnitude of the projection is greater than an acceptance threshold.

Then, to decide if a sub-window is an airplane, the image window must be projected over the subspace. The vector projection will be the product of each one of the basis vectors - eigenvectors - with the image window less the subspace mean image (figure 4.2). This is equal to make the product with the eigenvectors and subtract from the projected vector the subspace mean projection. The product of the mean subspace and the eigenvectors are zero, since each ring of the eigenvectors has zero mean and each ring of the subspace mean is constant. Then, as the eigenvectors are symmetric, the projection of all the sub-windows of the image can be directly computed convolving the eigenfilters with the image. These convolutions are computed in the Fourier domain taking advantage of the FFT. An image sub-window will be considered to be an airplane if the magnitude of its projection is large. The sum of the squares of the convolved images will give in each point the square of the magnitude of the projection of the window in the rotations orbit subspace. As the image subwindows have not exactly the same energy level, instead of the magnitude of the projection we have $\mathbf{E}^2$ times the magnitude of the projection, where $\mathbf{E}^2$ is the energy of the window. So that the response at each point must be divided by the 64×64 window energy centred at that point. The energy of each image window is easily computed convolving the square of the original image with a 64×64 window filled with ones.

Figure 4.1 (right) shows the square of the projection magnitude for the subspace defined with the selected eigenvectors. It can be seen how five of seven airplanes are correctly located. The encircled regions of the resultant image correspond to the five highest level "blobs" of the image. The reason that makes the system to fail on identifying all the airplanes is the size and background variation of the targets with respect to the model.

The efficiency improvement of using this kind of method must be seen separately for the recognition and learning stage. In the recognition stage, instead of convolving with the airplane images at each different orientation, it is performed more efficiently convolving with only a reduced number of eigenvectors. During the learning stage, the method supposes one scan of the image for the Cartesian to polar coordinates transformation, perform M fast Fourier transforms of a vector of dimension N for the computation of the eigenvalues and the scanning of the selected eigenvector (each one of $M \times N$ pixels). This is a more efficient process than computing the eigenvectors and eigenvalues with the conventional methods. Finally, The method permits the application of a pyramidal convolution approach so that it can be speeded up.

5. Conclusions

This chapter presents a method of rotational invariant pattern matching. The proposed method has been successfully applied to airplane detection in satellite images.

Further work will focus on the improvement of the developed method to avoid the response variation of the system due to differences of the image respect to the background of the target.

Acknowledgement. This work has been supported by grant TAP96-0629-c04-03 of the Comisión Interministerial de Ciencia y Tecnología (CICYT).

References

1. M. K. Hu, "Visual pattern recognition by moment invariants", *IEEE Transactions on Information Theory*, vol. 8, pp. 179-187, 1962.
2. I. T. Jollife, *Principal component analysis*, Springer-Verlag, New York, 1986.
3. M. Uenohara and T. Kanade, "Optimal approximation of uniformly rotated images: relationship between Karhunen-Loève expansion and discrete cosine transform", *IEEE Transactions on Image Processing*, vol. 7, no. 1, pp. 116-119, 1998.
4. T. Kohonen, *Self-organizing maps*, Springer Series in Information Sciences, Second Edition, 1997.
5. B. Moghaddam and A. Pentland, "Probabilistic visual learning for object representation", *IEEE Transactions on Pattern Analysis and Machine Intelligence*, vol. 19, no. 7, pp. 696-711, 1997.

6. J. Mundy and A. Zisserman, *Geometric invariance in computer vision*, The MIT Press, 1992.
7. H. Murase and S. Nayar, "Visual learning and recognition of 3-D objects from Appearance", *International Journal of Computer Vision*, vol. 14, pp. 5–24, 1995.
8. R. W. Picard, T. Kabir and F. Liu, "Real-time recognition with the entire Brodatz texture database", in *Proceedings of the IEEE Conference on Computer Vision and Pattern Recognition (CVPR'93)*, New York, pp. 638–639, June 1993.
9. A. Poularikas, *The transforms and applications handbook*, CRC Press, and IEEE Press, 1996.
10. A. Pujol, A. Solé, D. Ponsa, J. Varona and J. J. Villanueva, "Robust rotational invariant coin shape identification", in *Proceedings of the Workshop on European Scientific and Industrial Collaboration WSIC'98*, pp. 345–349, 1998.
11. K. R. Rao and P. Yip, *Discrete cosine transform, algorithms, advantages, applications*, Academic Press, 1990.
12. B. D. Ripley, *Pattern Recognition and Neural Networks*, Cambridge University Press, 1996.
13. H. Weschler, *Computational Vision*, Academic Press Inc., 1990.
14. J. Wood, "Invariant pattern recognition: a review", *Pattern Recognition*, vol. 29, no. 1, pp. 1–17, 1996.

Supervised Segmentation by Region Merging

Ben Gorte

ITC, Enschede, the Netherlands

Summary. A region merging segmentation method is combined with supervised classification to delineate and categorize objects in the terrain on the basis of remotely sensed imagery. The segmentation algorithm is a hybrid between region growing and split and merge. The algorithm makes a recursive, bottom up quadtree traversal, which starts at single pixels (or larger quadtree leaves in which the pixel values are constant) and recursively merges adjacent regions, forming irregularly shaped segments at all stages. Order dependency problems are solved by performing several iterations, while slowly relaxing the homogeneity criteria until a user defined degree of segmentation is reached.

By making the algorithm output segmentations at several threshold levels, a segmentation pyramid is created. Combining this with supervised classification allows for the selection of segments from different pyramid levels, which yields a partitioning of the space into objects, while preventing object fragmentation and object merging.

1. Introduction

Although image segmentation has received much attention in the computer vision literature since the 1970's [7], it has not become widely accepted in the field of remotely sensed image analysis. For example, in some very respected textbooks in this field image segmentation is not mentioned. Also most of the major commercial digital image processing software packages do not include image segmentation. Pavlidis [10] states that in spite of increased understanding of the nature of images we have been very slow in integrating the results of this understanding into useful image analysis programs. Nowadays, a number of problems are still associated with image segmentation. These have been summarized by [1] as: object merging, poor boundary localization, object boundary ambiguity, object fragmentation and sensitivity to noise.

The success of any segmentation algorithm depends on the availability of

- High resolution imagery, such that relevant objects are represented by a significant number of pixels; otherwise there is no point in segmentation.
- Powerful hardware: fast and with a large memory capacity.
- An efficient implementation, considering the sizes of remote sensing images.

Since the first two requirements are increasingly being fulfilled, it is worthwhile focusing on the third and to try to re-introduce image segmentation in Earth observation image analysis.

When image segmentation is combined with supervised classification, identification and labelling of terrain objects may be achieved. This chapter gives an algorithm for such an integration.

2. Image Segmentation

The purpose of image segmentation is to subdivide an image into regions that are homogeneous according to certain criteria, in such a way that these regions correspond to relevant objects in the terrain. The relevance of objects depends on user requirements.

A special case, which is typical for Earth observation applications, is multi-band imagery. Grey-scale segmentations ([2, 9]) of the individual bands do not exploit the full image information content. Each band gives a different set of segments, which creates additional difficulties when they are to be combined. A method is presented below that segments a multi-spectral image into one unique set of objects.

2.1 Existing Methods

The two major approaches in image segmentation are *edge based* and *region based*.

2.1.1 Edge Based Segmentation. Edge based segmentation is executed in two steps. The first step is to find segment boundaries in the image by identifying edge pixels, at those places where grey value changes occur. This is a neighbourhood operation: in order to decide whether a pixel is an edge pixel, neighbouring pixels have to be examined. Subsequently, each image region that is completely surrounded by edge pixels becomes a segment. A problem with this approach is that edge pixels, identified during the first step, do not obey topological constraints for segment boundaries. Therefore, an intermediate step is necessary to remove superfluous edge pixels and to fill gaps in boundaries. Edge based segmentation divides the image pixels into two kinds: those belonging to segments and those belonging to boundaries. This corresponds to a model for object representation in the raster domain where 'object pixels' are labeled with the object they belong to, and a separate label is reserved for 'boundary pixels' [8].

2.1.2 Region Based Segmentation. Area or region based segmentation creates segments by applying homogeneity criteria inside candidate segments. A distinction is made between *region growing* and *split and merge* algorithms. Region growing can be implemented in different ways, for example as follows:- Segments are formed starting from (randomly placed) seed pixels by iteratively augmenting them with surrounding pixels -as long as the homogeneity criteria are satisfied. When no more pixels can be attributed to any of the segments, new seeds are placed in the unsegmented areas and the process is repeated. This continues until the whole image is segmented. Split and merge algorithms start by subdividing the image into squares of a fixed size, usually corresponding to leaves at a certain level in a quadtree. Recursively, leaves are tested for homogeneity and heterogeneous leaves are subdivided into four lower level ones, while homogeneous leaves may be combined with three (homogeneous) neighbours into one leaf at a higher level, provided the homogeneity criteria continue to be satisfied. The recursion stops at the low end at single-pixel leaves (they are homogeneous), and at the high end when no further combinations can be made (the extreme case being an entirely homogeneous image). Subsequently, adjacent leaves at different levels are combined into irregularly shaped, homogeneous segments.

After region-based segmentation, each pixel belongs to a segment. There are no boundary pixels. This corresponds to the raster model which labels a pixel according to the object that has the largest overlap with the cell [8]. The advantage of this model, compared to the above-mentioned model that distinguishes between object and boundary pixels, is that the objects form a spatial partitioning. Also the terrain is usually regarded as being completely filled with objects. Moreover, as long as spatial resolutions are still an important limiting factor in satellite image applications, boundary pixels may completely obscure small objects. For example it is not clear how to represent a 10 m wide road in a 10 m resolution map using boundary pixels.

Region-based segmentations generally suffer from *order dependency*. During region growing, a segment could be expanded with any of a subset of neighbouring pixels, but not with all of them. Conversely, a pixel can be adjacent to more than one segment and might be added to each of those. The choices made in those cases are, to a certain degree, arbitrary and they are usually influenced by the order in which the data are stored and possible combinations examined. Similar considerations apply to the split and merge approach. In the initial (recursive) phase, the homogeneous regions are restricted by the quadtree structure. They are square, their sizes are powers of two, and they can only be located at a limited set of positions within the image. When merging leaves into segments during the final stage of split and merge, the order in which the combinations are examined plays a role.

2.2 Region Merging Segmentation

A *region merging* segmentation method was presented in [4], which does not show order dependency problems. The algorithm is based on connected component labeling in the region quadtree domain. It is a hybrid between region growing and split and merge. The algorithm makes a recursive, bottom up quadtree traversal, which starts at single pixels (or larger quadtree leaves in which the pixel values are constant) and recursively merges adjacent regions, forming irregularly shaped segments at all stages.

The coarseness of the segmentation is controlled by a merging criterion that involves a spectral distance threshold. Two adjacent (candidate) regions are merged if the distance between the average feature vectors of the pixels in each region does not exceed the threshold value. Moreover, merging only takes place if the variances and covariances over the feature vectors of all pixels in the combined region are smaller than the square of the threshold value.

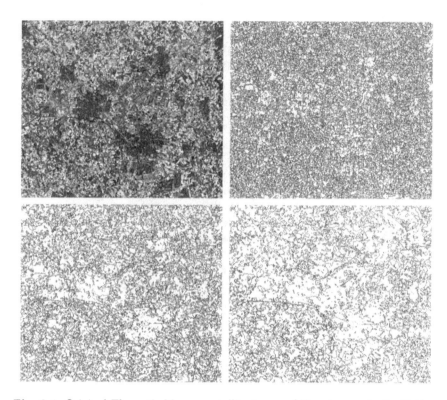

Fig. 2.1. Original Thematic Mapper satellite image of Twente area in the Netherlands, and segmentations with different thresholds.

Connected component labeling is a special case of this segmentation method, with a threshold equal to zero.

The order dependency problem is solved by performing several iterations, while gradually relaxing the homogeneity criteria (increasing the threshold value) until a user defined degree of segmentation is reached. During each iteration only a limited number of region-pairs satisfy the criteria and, therefore, arbitrary (i.e. order-dependent) choices are avoided. The iterative approach is feasible, because much attention was paid to an efficient implementation of the algorithm directly in the quadtree domain. The coarseness of the segmentation depends on the value of the threshold during the final iteration (figure 2.1)

When the final threshold value is large, a coarse segmentation is obtained, where segments are generally large and contain several terrain objects (merging). With a low final value, terrain objects that are not spectrally uniform are subdivided into several segments (fragmentation). Therefore, the user is confronted with the difficult task of finding an optimal threshold value for a compromise between these two extremes, which will still show examples of both fragmentation and merging. Due to spectral similarity of adjacent objects on the one hand and spectral heterogeneity within objects on the other, spectral information alone is usually not sufficient to delineate terrain objects in multi-spectral satellite imagery.

2.3 Extension to Segmentation Pyramids

A slightly modified version of the segmentation algorithm outputs segmentations at several threshold levels, thereby creating a segmentation pyramid. The levels of the pyramid (from top to bottom) represent segmentations with different degrees of coarseness (from coarse to fine). It is expected that terrain objects correspond to segments, which, however, reside at different pyramid levels.

The next section applies a refined supervised classification method to the segmentations in a pyramid, in order to select segments from different pyramid levels on the basis of class homogeneity. The additional information thus introduced is expected to provide a partitioning of the space into objects, while preventing object fragmentation and object merging. Moreover, each object will obtain a class label from a set of user-defined thematic classes, concerning for example land-cover or land-use.

3. Region Labeling

To select from a segmentation pyramid those segments that correspond to terrain objects, an adapted supervised image classification algorithm is used [6]. Given a user-defined set of classes $\{C_1 \ldots C_N\}$ with a number of training samples for each class (figure 3.1), the algorithm estimates the (relative) areas of these classes within each segment of an arbitrary segmentation of the image.

The method is based on standard Bayesian classification, which calculates the a posteriori probability $P(C_i|\mathbf{x})$ that class C_i should be assigned to a pixel with feature vector $\mathbf{x}$ using the Bayes formula:

$$P(C_i|\mathbf{x}) = \frac{P(\mathbf{x}|C_i) \, P(C_i)}{P(\mathbf{x})}$$

At the right hand side of this equation are the probability density $P(\mathbf{x}|C_i)$, which is estimated from training samples, the prior probability $P(C_i)$, which may be specified by the user on the basis of expected relative class areas, and the unconditional feature density $P(\mathbf{x})$, which is usually obtained by normalization. All the involved probabilities relate to the entire image, i.e. are independent of the position in the image. As a consequence, the probabilities for a particular pixel are influenced by statistics that also concern (possibly very different) regions elsewhere in the image.

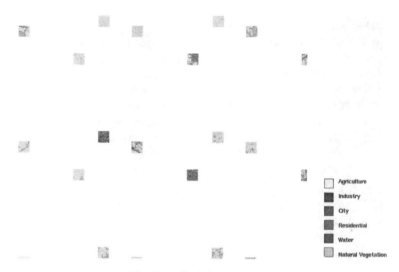

Fig. 3.1. Training samples.

3.1 Local Statistics

To refine the probability estimates, the concept of local statistics is introduced, where probabilities concern populations that are restricted to image segments, instead of the entire image.

To estimate probability densities, modified k-Nearest Neighbour estimation is applied, which finds for each feature vector $\mathbf{x}$ the classes of the k nearest neighbours in the feature space of training samples, which yields a sequence $k_i, i \in [1 \cdots N]$. The probability densities $P(\mathbf{x}|C_i)$ in classes C_i are proportional to k_i/T_i [3], where T_i is the total number of training samples for class C_i. The probability densities obtained in this way are *global*, because they only depend on $\mathbf{x}$ and not on the pixel position. *Local* probability densities, concerning a segment s, are obtained when T_i is replaced by T_i^s, the number of class C_i training samples involved in the classification of the segment that contains the pixel under consideration [5].

In [6] an iterative procedure is described to estimate local prior probabilities $P(C_i)$ on the basis of the image data, in combination with a collection of representative training samples. The procedure estimates the area A_i for class C_i in a segment s as the sum of the posterior probabilities $P(C_i|\mathbf{x}_p)$ over the pixels p in segments s, which contains A pixels:

$$A_i = \sum_{p=1}^{A} P(C_i|\mathbf{x}_p),$$

where, according to the Bayes formula

$$P(C_i|\mathbf{x}_p) = \frac{P(\mathbf{x}_p|C_i) \, P(C_i)}{P(\mathbf{x}_p)}.$$

Note that $A = \sum i = 1^N A_i$. The probability densities $P(\mathbf{x}_p|C_i)$ are estimated locally, as described above. The procedure is started with an initial set of equal prior probabilities $P(C_i) = 1/N, i \in [1 \cdots N]$. By normalizing A_i a new set is obtained

$$P(C_i) = A_i/A,$$

which is used in the next iteration. In [6] it is shown that the prior probabilities converge and that the result is correct if the probability density estimates are accurate.

On the basis of the final class area estimates $A_i, i \in [1 \cdots N]$ *pure* segments can be identified as those where the maximum A_i is equal (or almost equal) to A. The corresponding class label is assigned to such a segment. Segments that do not fulfill this condition are *mixed*.

4. Segment Selection

It is expected that many of the terrain objects are represented by corresponding segments, which, however, are residing at different levels of the pyramid. Therefore, the task becomes one of selecting a set of segments from the pyramid. This is done by applying region labelling to each segment at every level, and using the results to select those segments from the pyramid which represent relevant objects.

We select segments from different pyramid levels such that the union of those segments covers the entire area. Preferably, *pure* segments are selected, which, however do not yield a complete coverage of the area. Apparently, in certain regions, the classes cannot be spatially distinguished because of spatial or spectral resolution limitations. In those regions mixed segments are selected. If pure segments at different levels of the pyramid coincide, the one at the highest level has priority, because it has a larger area. Also the mixed segments, which fill the gaps between the pure ones, are selected as large as possible, i.e. from the highest possible pyramid levels (figure 4.1).

5. Conclusion

The chapter shows that region merging segmentation, especially when combined with supervised classification involving local statistics, provides a subdivision of the image into segments that are expected to coincide with terrain objects. The method is demonstrated using a Thematic Mapper satellite image with a spatial resolution of 30m. When higher resolution imagery becomes available, the approach will be more valuable, since the average number of pixels per terrain object will increase at a given mapping scale, which leads to an increased accuracy of probability estimates.

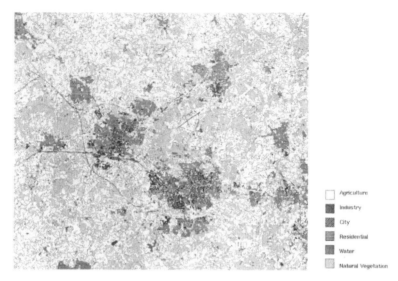

Fig. 4.1. Selected pure and mixed segments, labelled with (predominant) classes.

References

1. S. T. Acton, "On unsupervised segmentation of remotely sensed imagery using nonlinear regression", *International Journal of Remote Sensing*, vol. 17, no. 7, pp. 1407–1415, 1996.
2. Y.-L. Chang and X. Li, "Fast image region growing", *Image and Vision Computing*, vol. 13, pp. 559–571, 1995.
3. R. O. Duda and P. E. Hart, *Pattern classification and scene analysis*. John Wiley & Sons, New York, 1973.
4. B. G. H. Gorte, "Multi-spectral quadtree based image segmentation", in *Proceedings, 18th International Society for Photogrammetry and Remote Sensing Congress*, Vienna, ISPRS XXXI, Part B3, Comm.III, pp. 251–256, 1996.
5. B. G. H. Gorte, "Local statistics in supervised classification", in *Proceedings of European Association of Remote Sensing Laboratories, (EARSEL), Operational Remote Sensing for Sustainable Development*, Enschede, NL, 11–15 May 1998 (in press).
6. B. G. H. Gorte and A. Stein, "Bayesian classification and class area estimation of Satellite Images using Stratification", *IEEE Transactions on Geoscience and Remote Sensing*, vol. 36, no. 3, May 1998.
7. S. L. Horowitz and T. Pavlidis, "Picture segmentation by a tree traversal algorithm", *Journal of the Association of Computing Machinery*, vol. 23, pp. 368–388, 1976.
8. M. Molenaar. *An introduction to the theory of spatial object modelling for GIS*, 1998 (in press, preliminary version 233 pp.).
9. O. J. Morris, M. deJ. Lee, A .G. Constantinides, "Graph theory for image analysis: an approach based on the shortest spanning tree", *IEE Proceedings-Part F*, vol. 133, pp. 146–152, 1986.
10. T. Pavlidis. "A critical survey of image analysis methods", *IAPR-8*, pp. 502–511, 1986.

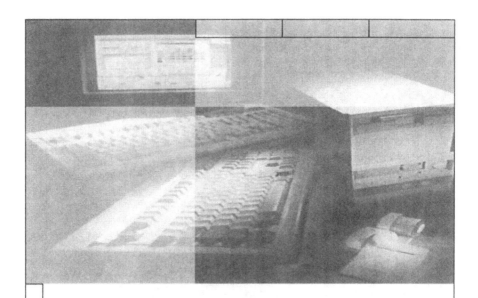

I. Kanellopoulos, G.G. Wilkinson,
F. Roli, J. Austin (Eds.)

Neurocomputation in Remote Sensing Data Analysis

Proceedings of Concerted Actions "COMPARES" (Connectionist Methods for Pre-Processing and Analysis of Remote Sensing Data)

1997. IX, 284 pp. 87 figs., 39 tabs.
Hardcover **DM 148,-***;
öS 1081,-; sFr 135,-
ISBN 3-540-63316-2

This volume gives a state of the art view of recent developments in the use of artificial neural networks for the analysis of remotely sensed satellite data. Remote sensing has now become a discipline in which ever increasing volumes of data, gathered from space together with growing application needs for high precision spatial products, need to be interpreted in shorter times and with in-creasing accuracy. Neural networks, as a new form of computational paradigm, seem well suited to many of the tasks involved in remotely sensed image analysis. This book demonstrates a wide range of uses of neural networks for remote sensing applications and provides the views of a large number European experts brought together in the framework of a concerted action supported by the European Commission.

*This price applies in Germany/Austria/Switzerland and is a recommended retail price.
Prices and other details are subject to change without notice.
In EU countries the local VAT is effective.

Springer-Verlag · Postfach 14 02 01 · D-14302 Berlin
Tel.: 0 30 / 82 787 - 2 32 · http://www.springer.de
Bücherservice: Fax 0 30 / 82 787 - 3 01
e-mail: orders@springer.de

d&p · 65571/1 SF

Springer and the environment

At Springer we firmly believe that an international science publisher has a special obligation to the environment, and our corporate policies consistently reflect this conviction.
We also expect our business partners – paper mills, printers, packaging manufacturers, etc. – to commit themselves to using materials and production processes that do not harm the environment. The paper in this book is made from low- or no-chlorine pulp and is acid free, in conformance with international standards for paper permanency.

CPSIA information can be obtained at www.ICGtesting.com
Printed in the USA
LVOW072159050912

297591LV00002B/68/P